Data Quality and Record
Linkage Techniques

Thomas N. Herzog
Fritz J. Scheuren
William E. Winkler

Data Quality and Record Linkage Techniques

 Springer

Thomas N. Herzog
Office of Evaluation
Federal Housing Administration
U.S. Department of Housing and Urban Development
451 7-th Street, SW
Washington, DC 20140

Fritz J. Scheuren
National Opinion Research Center
University of Chicago
1402 Ruffner Road
Alexandria, VA 22302

William E. Winkler
Statistical Research Division
U.S. Census Bureau
4700 Silver Hill Road
Washington, DC 20233

Library of Congress Control Number: 2007921194

ISBN-13: 978-0-387-69502-0   e-ISBN-13: 978-0-387-69505-1

Printed on acid-free paper.

Preface

Readers will find this book a mixture of practical advice, mathematical rigor, management insight, and philosophy. Our intended audience is the working analyst. Our approach is to work by real life examples. Most illustrations come out of our successful practice. A few are contrived to make a point. Sometimes they come out of failed experience, ours and others.

We have written this book to help the reader gain a deeper understanding, at an applied level, of the issues involved in improving data quality through editing, imputation, and record linkage. We hope that the bulk of the material is easily accessible to most readers although some of it does require a background in statistics equivalent to a 1-year course in mathematical statistics. Readers who are less comfortable with statistical methods might want to omit Section 8.5, Chapter 9, and Section 18.6 on first reading. In addition, Chapter 7 may be primarily of interest to those whose professional focus is on sample surveys. We provide a long list of references at the end of the book so that those wishing to delve more deeply into the subjects discussed here can do so.

Basic editing techniques are discussed in Chapter 5, with more advanced editing and imputation techniques being the topic of Chapter 7. Chapter 14 illustrates some of the basic techniques. Chapter 8 is the essence of our material on record linkage. In Chapter 9, we describe computational techniques for implementing the models of Chapter 8. Chapters 9–13 contain techniques that may enhance the record linkage process. In Chapters 15–17, we describe a wide variety of applications of record linkage. Chapter 18 is our chapter on data confidentiality, while Chapter 19 is concerned with record linkage software. Chapter 20 is our summary chapter.

Three recent books on data quality – Redman [1996], English [1999], and Loshin [2001] – are particularly useful in effectively dealing with many management issues associated with the use of data and provide an instructive overview of the costs of some of the errors that occur in representative databases. Using as their starting point the work of quality pioneers such as Deming, Ishakawa, and Juran whose original focus was on manufacturing processes, the recent books cover two important topics not discussed by those seminal authors: (1) errors that affect data quality even when the underlying processes are operating properly and (2) processes that are controlled by others (e.g., other organizational units within one's company or other companies).

Dasu and Johnson [2003] provide an overview of some statistical summaries and other conditions that must exist for a database to be useable for specific statistical purposes. They also summarize some methods from the database literature that can be used to preserve the integrity and quality of a database. Two other interesting books on data quality – Huang, Wang and Lee [1999] and Wang, Ziad, and Lee [2001] – supplement our discussion. Readers will find further useful references in The International Monetary Fund's (IMF) Data Quality Reference Site on the Internet at http://dsbb.imf.org/Applications/web/dqrs/dqrshome/.

We realize that organizations attempting to improve the quality of the data within their key databases do best when the top management of the organization is leading the way and is totally committed to such efforts. This is discussed in many books on management. See, for example, Deming [1986], Juran and Godfrey [1999], or Redman [1996]. Nevertheless, even in organizations not committed to making major advances, analysts can still use the tools described here to make substantial quality improvement.

A working title of this book – *Playing with Matches* – was meant to warn readers of the danger of data handling techniques such as editing, imputation, and record linkage unless they are tightly controlled, measurable, and as transparent as possible. Over-editing typically occurs unless there is a way to measure the costs and benefits of additional editing; imputation always adds uncertainty; and errors resulting from the record linkage process, however small, need to be taken into account during future uses of the data.

We would like to thank the following people for their support and encouragement in writing this text: Martha Aliaga, Patrick Ball, Max Brandstetter, Linda Del Bene, William Dollarhide, Mary Goulet, Barry I. Graubard, Nancy J. Kirkendall, Susan Lehmann, Sam Phillips, Stephanie A. Smith, Steven Sullivan, and Gerald I. Webber.

We would especially like to thank the following people for their support and encouragement as well as for writing various parts of the text: Patrick Baier, Charles D. Day, William J. Eilerman, Bertram M. Kestenbaum, Michael D. Larsen, Kevin J. Pledge, Scott Schumacher, and Felicity Skidmore.

Contents

x   Contents

About the Authors

Thomas N. Herzog, Ph.D., ASA, is the Chief Actuary at the US Department of Housing and Urban Development. He holds a Ph.D. in mathematics from the University of Maryland and is also an Associate of the Society of Actuaries. He is the author or co-author of books on Credibility Theory, Monte Carlo Methods, and Risk Models. He has devoted a major effort to improving the quality of the databases of the Federal Housing Administration.

Fritz J. Scheuren, Ph.D., is a general manager with the National Opinion Research Center. He has a Ph.D. in statistics from the George Washington University. He is much published with over 300 papers and monographs. He is the 100th President of the American Statistical Association and a Fellow of both the American Statistical Association and the American Association for the Advancement of Science. He has a wide range of experience in all aspects of survey sampling, including data editing and handling missing data. Much of his professional life has been spent employing large operational databases, whose incoming quality was only marginally under the control of the data analysts under his direction. His extensive work in recent years on human rights data collection and analysis, often under very adverse circumstances, has given him a clear sense of how to balance speed and analytic power within a framework of what is feasible.

William E. Winkler, Ph.D., is Principal Researcher at the US Census Bureau. He holds a Ph.D. in probability theory from Ohio State University and is a fellow of the American Statistical Association. He has more than 110 papers in areas such as automated record linkage and data quality. He is the author or co-author of eight generalized software systems, some of which are used for production in the largest survey and administrative list situations.

1
Introduction

1.1. Audience and Objective

This book is a primer on editing, imputation and record linkage for analysts who are responsible for the quality of large databases, including those sometimes known as data warehouses. Our goal is to provide practical help to people who need to make informed and cost-effective judgments about how and when to take steps to safeguard or improve the quality of the data for which they are responsible. We are writing for people whose professional day-to-day lives are governed, or should be, by data quality issues. Such readers are in academia, government, and the private sector. They include actuaries, economists, statisticians, and computer scientists. They may be end users of the data, but they are more often working in the middle of a data system. We are motivated to write the book by hard experience in our own working lives, where unanticipated data quality problems have cost our employers and us dearly–in both time and money. Such problems can even damage an organization's reputation.

Since most readers are familiar, at some level, with much of the material we cover, we do not expect, or recommend, that everyone read this book thoroughly from cover to cover. We have tried to be comprehensive, however, so that readers who need a brief refresher course on any particular issue or technique we discuss can get one without going elsewhere.

To be as user-friendly as possible for our audience, we mix mathematical rigor with practical advice, management insight, and even philosophy. A major point to which we return many times is the need to have a good understanding of the primary intended uses for the database, even if you are not an end user yourself.

1.2. Scope

Our goal is to describe techniques that the analyst can use herself or himself in three main areas of application:

(1) *To improve the useful quality of existing or contemplated databases/lists.*
   Here our aim is to describe analytical techniques that facilitate the

improvement of the quality of individual data items within databases/lists. This is the topic of the classic text of Naus [1975] – alas, now out-of-print. This first area of interest also entails describing techniques that facilitate the elimination of duplicate records from databases/lists.

(2) *To merge two or more lists.* The merging of two or more lists involves record linkage, our second area of study. Here the classic text is Newcombe [1988] – another book out-of-print.[1] Such lists may be mailing lists of retail customers of a large chain of stores. Alternatively, the merged list might be used as a sampling frame to select individual entities (e.g., a probability sample of farms in the United States) to be included in a sample survey. In addition to seeking a list that contains every individual or entity in the population of interest, we also want to avoid duplicate entries. We can use record linkage techniques to help us do this.

(3) *To merge two or more distinct databases.* The merging of two or more databases is done to create a new database that has more data elements than any previously existing (single) database, typically to conduct research studies. A simple example of this is a recent study (see Section 3.4 for further discussion) that merged a database of records on licensed airplane pilots with a database of records on individuals receiving disability benefits from the US Social Security Administration. The classic paper on this type of record linkage study is Newcombe et al. [1959]. Early applications of record linkage frequently focused on health and genetics issues.

1.3.   Structure

Our text consists of four parts.

Part I (Data Quality: What It Is, Why It Is Important, and How to Achieve It) consists of four chapters. In Chapter 2 we pose three fundamental questions about data quality that help in assessing a database's overall fitness for use. We use a systems perspective that includes all stages, from the generation of the initial datasets and ways to prevent errors from arising, to the data processing steps that take the data from data capture and cleaning, to data interpretation and analysis. In Chapter 3 we present a number of brief examples to illustrate the enormous consequences of successes–and failures–in data and database use. In Chapter 4, we describe metrics that quantify the quality of databases and data lists. In Chapter 5 we revisit a number of data quality control and editing techniques described in Naus [1975] and add more recent material that supplements his work in this area. Chapter 5 also includes a number of examples illustrating the techniques we describe.

---

[1] The fact that the classic texts for our first two areas of interest are both out of print and out of date is a major motivation for our book, which combines the two areas and brings together the best of these earlier works with the considerable methodological advances found in journals and in published and unpublished case studies since these two classic texts were written.

Part II of the text (Mathematical Tools for Editing, Imputation, and Record Linkage) is the book's heart and essence. Chapter 6 presents some mathematical preliminaries that are necessary for understanding the material that follows. In Chapter 7 we present an in-depth introductory discussion of specialized editing and imputation techniques within a survey-sampling environment. Our discussion of editing in sample surveys summarizes the work of Fellegi and Holt [1976]. Similarly, our treatment of imputation of missing data in sample surveys highlights the material of Rubin [1987] and Little and Rubin [2002]–books that provide an excellent treatment of that topic. Those dealing with non-survey data (e.g., corporate mailing lists or billing systems) frequently decide to use alternative schemes that devote more resources to ensuring that the items within their databases are correct than the data-correction techniques we concentrate on here. In Chapters 8 and 9, we describe the fundamental approaches to record linkage as presented by Fellegi–Sunter and Belin–Rubin. In Chapters 10–13, we describe other techniques that can be used to enhance these record linkage models. These include standardization and parsing (Chapter 10), phonetic coding systems for names (Chapter 11), blocking (Chapter 12), and string comparator metrics for typographical errors (Chapter 13).

In Part III (Case Studies on Record Linkage) we present a wide variety of examples to illustrate the multiple uses of record linkage techniques. Chapter 14 describes a variety of applications based on HUD's FHA single-family mortgage records. Other topics considered in Part III include medical, biomedical, highway safety, and social security.

In the last part of the text (Part IV) we discuss record linkage software and privacy issues relating to record linkage applications.

Part 1
Data Quality: What It Is, Why It Is Important, and How to Achieve It

2
What Is Data Quality and Why Should We Care?

Caring about data quality is key to safeguarding and improving it. As stated, this sounds like a very obvious proposition. But can we, as the expression goes, "recognize it when we see it"? Considerable analysis and much experience make it clear that the answer is "no." Discovering whether data are of acceptable quality is a measurement task, and not a very easy one. This observation becomes all the more important in this information age, when explicit and meticulous attention to data is of growing importance if information is not to become misinformation.

This chapter provides foundational material for the specifics that follow in later chapters about ways to safeguard and improve data quality.[1] After identifying when data are of high quality, we give reasons why we should care about data quality and discuss how one can obtain high-quality data.

Experts on quality (such as Redman [1996], English [1999], and Loshin [2001]) have been able to show companies how to improve their processes by first understanding the basic procedures the companies use and then showing new ways to collect and analyze quantitative data about those procedures in order to improve them. Here, we take as our primary starting point primarily the work of Deming, Juran, and Ishakawa.

2.1. When Are Data of High Quality?

Data are of high quality if they are "Fit for Use" in their intended operational, decision-making and other roles.[2] In many settings, especially for intermediate products, it is also convenient to define quality as "Conformance to Standards" that have been set, so that fitness for use is achieved. These two criteria link the

---

[1] It is well recognized that quality must have undoubted top priority in every organization. As Juran and Godfrey [1999; pages 4–20, 4–21, and 34–9] makes clear, quality has several dimensions, including meeting customer needs, protecting human safety, and protecting the environment. We restrict our attention to the quality of data, which can affect efforts to achieve quality in all three of these overall quality dimensions.
[2] Juran and Godfrey [1999].

role of the employee doing the work (conformance to standards) to the client receiving the product (fitness for use). When used together, these two can yield efficient systems that achieve the desired accuracy level or other specified quality attributes.

Unfortunately, the data of many organizations do not meet either of these criteria. As the cost of computers and computer storage has plunged over the last 50 or 60 years, the number of databases has skyrocketed. With the wide availability of sophisticated statistical software and many well-trained data analysts, there is a keen desire to analyze such databases in-depth. Unfortunately, after they begin their efforts, many data analysts realize that their data are too messy to analyze without major data cleansing.

Currently, the only widely recognized properties of quality are quite general and cannot typically be used without further elaboration to describe specific properties of databases that might affect analyses and modeling. The seven most commonly cited properties are (1) relevance, (2) accuracy, (3) timeliness, (4) accessibility and clarity of results, (5) comparability, (6) coherence, and (7) completeness.[3] For this book, we are primarily concerned with five of these properties: relevance, accuracy, timeliness, comparability, and completeness.

2.1.1. Relevance

Several facets are important to the relevance of the data analysts' use of data.

- Do the data meet the basic needs for which they were collected, placed in a database, and used?
- Can the data be used for additional purposes (e.g., a market analysis)? If the data cannot presently be used for such purposes, how much time and expense would be needed to add the additional features?
- Is it possible to use a database for several different purposes? A secondary (or possibly primary) use of a database may be better for determining what subsets of customers are more likely to purchase certain products and what types of advertisements or e-mails may be more successful with different groups of customers.

2.1.2. Accuracy

We cannot afford to protect against all errors in every field of our database. What are likely to be the main variables of interest in our database? How accurate do our data need to be?

---

[3] Haworth and Martin [2001], Brackstone [2001], Kalton [2001], and Scheuren [2001]. Other sources (Redman [1996], Wang [1998], Pipino, Lee, and Wang [2002]) provide alternative lists of properties that are somewhat similar to these.

For example, how accurate do our data need to be to predict:

- Which customers will buy certain products in a grocery store? Which customers bought products (1) this week, (2) 12 months ago, and (3) 24 months ago? Should certain products be eliminated or added based on sales trends? Which products are the most profitable?
- How will people vote in a Congressional election? We might be interested in demographic variables on individual voters – for example, age, education level, and income level. Is it acceptable here if the value of the income variable is within 20% of its true value? How accurate must the level of education variable be?
- How likely are individuals to die from a certain disease? Here the context might be a clinical trial in which we are testing the efficacy of a new drug. The data fields of interest might include the dosage level, the patient's age, a measure of the patient's general health, and the location of the patient's residence. How accurate does the measurement of the dosage level need to be? What other factors need to be measured (such as other drug use or general health level) because they might mitigate the efficacy of the new drug? Are all data fields being measured with sufficient accuracy to build a model to reliably predict the efficacy of various dosage levels of the new drug?

Are more stringent quality criteria needed for financial data than are needed for administrative or survey data?

2.1.3. Timeliness

How current does the information need to be to predict which subsets of customers are more likely to purchase certain products? How current do public opinion polls need to be to accurately predict election results? If data editing delays the publication/release of survey results to the public, how do the delays affect the use of the data in (1) general-circulation publications and (2) research studies of the resulting micro-data files?

2.1.4. Comparability

Is it appropriate to combine several databases into a data warehouse to facilitate the data's use in (1) exploratory analyses, (2) modeling, or (3) statistical estimation? Are data fields (e.g., Social Security Numbers) present within these databases that allow us to easily link individuals across the databases? How accurate are these identifying fields? If each of two distinct linkable databases[4] has an income variable, then which income variable is better to use, or is there a way to incorporate both into a model?

---

[4] This is illustrated in the case studies of the 1973 SSA-IRS-CPS exact match files discussed in Section 17.3 of this work.

2.1.5. Completeness

Here, by *completeness* we mean that no records are missing and that no records have missing data elements. In the survey sampling literature, entire missing records are known as *unit non-response* and missing items are referred to as *item non-response*. Both *unit* non-response and *item* non-response can indicate lack of quality. In many databases such as financial databases, missing entire records can have disastrous consequences. In survey and administrative databases, missing records can have serious consequences if they are associated with large companies or with a large proportion of employees in one subsection of a company. When such problems arise, the processes that create the database must be examined to determine whether (1) certain individuals need additional training in use of the software, (2) the software is not sufficiently user-friendly and responsive, or (3) certain procedures for updating the database are insufficient or in error.

2.2. Why Care About Data Quality?

Data quality is important to business and government for a number of obvious reasons. First, a reputation for world-class quality is profitable, a "business maker." As the examples of Section 3.1 show, high-quality data can be a major business asset, a unique source of competitive advantage.

By the same token, poor-quality data can reduce customer satisfaction. Poor-quality data can lower employee job satisfaction too, leading to excessive turnover and the resulting loss of key process knowledge. Poor-quality data can also breed organizational mistrust and make it hard to mount efforts that lead to needed improvements.

Further, poor-quality data can distort key corporate financial data; in the extreme, this can make it impossible to determine the financial condition of a business. The prominence of data quality issues in corporate governance has become even greater with enactment of the Sarbanes–Oxley legislation that holds senior corporate management responsible for the quality of its company's data.

High-quality data are also important to all levels of government. Certainly the military needs high-quality data for all of its operations, especially its counter-terrorism efforts. At the local level, high-quality data are needed so that individuals' residences are assessed accurately for real estate tax purposes.

The August 2003 issue of *The Newsmonthly of the American Academy of Actuaries* reports that the National Association of Insurance Commissioners (NAIC) suggests that actuaries audit "controls related to the completeness, accuracy, and classification of loss data". This is because poor data quality can make it impossible for an insurance company to obtain an accurate estimate of its insurance-in-force. As a consequence, it may miscalculate both its premium income and the amount of its loss reserve required for future insurance claims.

2.3. How Do You Obtain High-Quality Data?

In this section, we discuss three ways to obtain high-quality data.

2.3.1. *Prevention: Keep Bad Data Out of the Database/List*

The first, and preferable, way is to ensure that all data entering the database/list are of high quality. One thing that helps in this regard is a system that edits data before they are permitted to enter the database/list. Chapter 5 describes a number of general techniques that may be of use in this regard. Moreover, as Granquist and Kovar [1977] suggest, "The role of editing needs to be re-examined, and more emphasis placed on using editing to learn about the data collection process, in order to concentrate on preventing errors rather than fixing them."

Of course, there are other ways besides editing to improve the quality of data. Here organizations should encourage their staffs to examine a wide variety of methods for improving the entire process. Although this topic is outside the scope of our work, we mention two methods in passing. One way in a survey-sampling environment is to improve the data collection instrument, for example, the survey questionnaire. Another is to improve the methods of data acquisition, for example, to devise better ways to collect data from those who initially refuse to supply data in a sample survey.

2.3.2. *Detection: Proactively Look for Bad Data Already Entered*

The second scheme is for the data analyst to proactively look for data quality problems and then correct the problems. Under this approach, the data analyst needs at least a basic understanding of (1) the subject matter, (2) the structure of the database/list, and (3) methodologies that she might use to analyze the data. Of course, even a proactive approach is tantamount to admitting that we are too busy mopping up the floor to turn off the water.

If we have quantitative or count data, there are a variety of elementary methods, such as univariate frequency counts or two-way tabulations, that we can use. More sophisticated methods involve Exploratory Data Analysis (EDA) techniques. These methods, as described in Tukey [1977], Mosteller and Tukey [1977], Velleman and Hoaglin [1981], and Cleveland [1994], are often useful in examining (1) relationships among two or more variables or (2) aggregates. They can be used to identify anomalous data that may be erroneous.

Record linkage techniques can also be used to identify erroneous data. An extended example of such an application involving a database of mortgages is presented in Chapter 14. Record linkage can also be used to improve the quality of a database by linking two or more databases, as illustrated in the following example.

Example 2.1: Improving Data Quality through Record Linkage

Suppose two databases had information on the employees of a company. Suppose one of the databases had highly reliable data on the home addresses of the employees but only sketchy data on the salary history on these employees while the second database had essentially complete and accurate data on the salary history of the employees. Records in the two databases could be linked and the salary history from the second database could be used to replace the salary history on the first database, thereby improving the data quality of the first database.

2.3.3. Repair: Let the Bad Data Find You and Then Fix Things

By far, the worst approach is to wait for data quality problems to surface on their own. Does a chain of grocery stores really want its retail customers doing its data quality work by telling store managers that the scanned price of their can of soup is higher than the price posted on the shelf? Will a potential customer be upset if a price higher than the one advertised appears in the price field during checkout at a website? Will an insured whose chiropractic charges are fully covered be happy if his health insurance company denies a claim because the insurer classified his health provider as a physical therapist instead of a chiropractor? Data quality problems can also produce unrealistic or noticeably strange answers in statistical analysis and estimation. This can cause the analyst to spend lots of time trying to identify the underlying problem.

2.3.4. Allocating Resources – How Much for Prevention, Detection, and Repair

The question arises as to how best to allocate the limited resources available for a sample survey, an analytical study, or an administrative database/list. The typical mix of resources devoted to these three activities in the United States tends to be on the order of:

Prevent: 10%
Detect: 30%
Repair: 60%.

Our experience strongly suggests that a more cost-effective strategy is to devote a larger proportion of the available resources to preventing bad data from getting into the system and less to detecting and repairing (i.e., correcting) erroneous data. It is usually less expensive to find and correct errors early in the process than it is in the later stages. So, in our judgment, a much better mix of resources would be:

Prevent: 45%
Detect: 30%
Repair: 25%.

2.4.   Practical Tips

2.4.1.   Process Improvement

One process improvement would be for each company to have a few individuals who have learned additional ways of looking at available procedures and data that might be promising in the quest for process improvement. In all situations, of course, any such procedures should be at least crudely quantified – before adoption – as to their potential effectiveness in reducing costs, improving customer service, and allowing new marketing opportunities.

2.4.2.   Training Staff

Many companies and organizations may have created their procedures to meet a few day-to-day processing needs, leaving them unaware of other procedures for improving their data. Sometimes, suitable training in software development and basic clerical tasks associated with customer relations may be helpful in this regard. Under other conditions, the staff members creating the databases may need to be taught basic schemes for ensuring minimally acceptable data quality.

   In all situations, the company should record the completion of employee training in appropriate databases and, if resources permit, track the effect of the training on job performance. A more drastic approach is to obtain external hires with experience/expertise in (1) designing databases, (2) analyzing the data as they come in, and (3) ensuring that the quality of the data produced in similar types of databases is "fit for use."

2.5.   Where Are We Now?

We are still at an early stage in our discussion of data quality concepts. So, an example of what is needed to make data "fit for use" might be helpful before continuing.

Example 2.2: Making a database fit for use

Goal: A department store plans to construct a database that has a software interface that allows customer name, address, telephone number and order information to be collected accurately.

Developing System Requirements: All of the organizational units within the department store need to be involved in this process so that their operational needs can be met. For instance, the marketing department should inform the database designer that it needs both (1) a field indicating the amount of money each customer spent at the store during the previous 12 months and (2) a field indicating the date of each customer's most recent purchase at the store.

Data Handling Procedures: Whatever procedures are agreed upon, clear instructions must be communicated to all of the affected parties within the department store. For example, clear instructions need to be provided on how to handle missing data items. Often, this will enable those maintaining the database to use their limited resources most effectively and thereby lead to a higher quality database.

Developing User Requirements – How will the data be used and by whom? All of the organizational units within the department store who expect to use the data should be involved in this process so that their operational needs can be met. For example, each unit should be asked what information they will need. Answers could include the name and home address for catalog mailings and billing, an e-mail address for sale alerts, and telephone number(s) for customer service. How many phone numbers will be stored for each customer? Three? One each for home, office, and mobile? How will data be captured? Are there legacy data to import from a predecessor database? Who will enter new data? Who will need data and in what format? Who will be responsible for the database? Who will be allowed to modify data, and when? The answers to all these questions impact the five aspects of quality that are of concern to us.

Relevance: There may be many uses for this database. The assurance that all units who could benefit from using the data do so is one aspect of the relevance of the data. One thing that helps in this regard is to make it easy for the store's employees to access the data. In addition, addresses could be standardized (see Chapter 10) to facilitate the generation of mailing labels.

Accuracy: Incorrect telephone numbers, addresses, or misspelled names can make it difficult for the store to contact its customers, making entries in the database of little use. *Data editing* is an important tool for finding errors, and more importantly for ensuring that only correct data enter the system at the time of data capture. For example, when data in place name, state, and Zip Code fields are entered or changed, such data could be subjected to an edit that ensures that the place name and Zip Code are consistent. More ambitiously, the street address could be parsed (see Chapter 10) and the street name checked for validity against a list of the streets in the city or town. If legacy data are to be imported, then they should be checked for accuracy, timeliness, and duplication before being entered into the database.

Timeliness: Current data are critical in this application. Here again, *record linkage* might be used together with external mailing lists, to confirm the customers' addresses and telephone numbers. Inconsistencies could be resolved in order to keep contact information current. Further, procedures such as real-time data capture (with editing at the time of capture) at the first point of contact with the customer would allow the database to be updated exactly when the customer is acquired.

Comparability: The database should capture information that allows the department store to associate its data with data in its other databases (e.g., a trans-actions database). Specifically, the store wants to capture the names, addresses, and telephone numbers of its customers in a manner that enables it to link its customers across its various databases.

Completeness: The department store wants its database to be complete, but customers may not be willing to provide all of the information requested. For example, a customer may not wish to provide her telephone number. Can these missing data be obtained, or *imputed*, from public sources? Can a nine-digit Zip Code be imputed from a five-digit Zip Code and a street address? (Anyone who receives mail at home knows that this is done all the time.) Can a home telephone number be obtained from the Internet based on the name and/or home address? What standard operating procedures can be established to ensure that contact data are obtained from every customer? Finally, *record linkage* can be used to eliminate duplicate records that might result in a failure to contact a customer, or a customer being burdened by multiple contacts on the same subject.

This simple example shows how the tools discussed in this book – data editing, imputation, and record linkage – can be used to improve the quality of data. As the reader will see, these tools grow in importance as applications increase in complexity.

3
Examples of Entities Using Data to their Advantage/Disadvantage

In this chapter we first summarize work at five companies that are making successful use of their data for competitive advantage. We then discuss problems with the quality of some other databases and the resulting disastrous consequences. Our next example illustrates fitness-for-use complexities by considering efforts to obtain reliable data on the ages of the oldest residents of several countries. We conclude with (1) a dramatic example of the effectiveness of matching case records between two distinct databases and (2) a brief discussion of the use of a billing system within a medical practice.

3.1. Data Quality as a Competitive Advantage

In the five examples that follow, we show how relevance, accuracy, timeliness, comparability, and completeness of data can yield competitive advantage.

3.1.1. Harrah's[1]

Harrah's, a hotel chain that features casinos, collects lots of data on its customers – both big-time spenders and small, but steady gamblers – through its Total Rewards Program®. At the end of calendar year 2003, the Total Rewards Program included 26 million members of whom 6 million had used their membership during the prior 12 months.

The database for this program is based on an integrated, nationwide computer system that permits real-time communication among all of Harrah's properties. Harrah's uses these data to learn as much as it can about its customers in order to give its hotel/casino guests customized treatment. This enables Harrah's to know the gambling, eating, and spending preferences of its customers. Hence, Harrah's

---

[1] This section is based in part on Jill Griffin's article "How Customer Information Gives Harrah's a Winning Hand" that can be found at http://www.refresher.com/ !jlgharrahs.html.

can tailor its services to its customers by giving customized complimentary services such as free dinners, hotel rooms, show tickets, and spa services.

While the prevailing wisdom in the hotel business is that the attractiveness of a property drives business, Harrah's further stimulates demand by knowing its customers. This shows that Harrah's is listening to its customers and helps Harrah's to build customer loyalty.

Harrah's has found that this increased customer loyalty results in more frequent customer visits to its hotels/casinos with a corresponding increase in customer spending. In fact, according to its *2004 Annual Report* Harrah's "believes that its portion of the customer gaming budget has climbed from 36 percent in 1998 to more than 43 percent" in 2002.

3.1.2. Wal-Mart

According to Wal-Mart's *2005 Annual Report*, Wal-Mart employs over 75,000 people in Logistics and in its Information Systems Division. These employees enable Wal-Mart to successfully implement a "retailing strategy that strives to have what the customer wants, when the customer wants it."

With the Data Warehouse storage capacity of over 570 terabytes – larger than all of the fixed pages on the internet – we [Wal-Mart] have [put] a remarkable level of real-time visibility planning into our merchandise planning. So much so that when Hurricane Ivan was heading toward the Florida panhandle, we knew that there would be a rise in demand for Kellogg's® Strawberry Pop-Tart® toaster pastries. Thanks to our associates in the distribution centers and our drivers on the road, merchandise arrived quickly.

3.1.3. Federal Express[2]

FedEx InSight is a real-time computer system that permits Federal Express' business customers to go on-line to obtain up-to-date information on all of their Federal Express cargo information. This includes outgoing, incoming, and third-party[3] shipments. The business customer can tailor views and drill down into freight information, including shipping date, weight, contents, expected delivery date, and related shipments. Customers can even request e-mail notifications of in-transit events, such as attempted deliveries and delays at customs and elsewhere.

InSight links shipper and receiver data on shipping bills with entries in a database of registered InSight customers. The linking software, developed by Trillium Software, is able to recognize, interpret, and match customer names and address information. The challenge in matching records was not with the records of outgoing shippers, who could be easily identified by their account number.

---

[2] This is based on http://www.netpartners.com.my/PDF/Trillium%20Software%20Case% 20Study%20%20FedEx.pdf.
[3] For example, John Smith might buy a gift from Amazon.com for Mary Jones and want to find out if the gift has already been delivered to Mary's home.

The real challenge was to link the intended shipment recipients to customers in the InSight database.

The process, of course, required accurate names and addresses. The addresses on the Federal Express bills tend not to be standardized and to be fraught with errors, omissions, and other anomalies. The bills also contain a lot of extraneous information such as parts numbers, stock keeping units, signature requirements, shipping contents, and delivery instructions. These items make it harder to extract the required name and address from the bill. The system Trillium Software developed successfully met all of these challenges and was able to identify and resolve matches in less than 1 second, processing as many as 500,000 records per hour.

3.1.4.   Albertsons, Inc. (and RxHub)

Albertsons is concerned with the safety of the customers buying prescription drugs at its 1900 pharmacies. It is crucial that Albertsons correctly identify all such customers. Albertsons needs an up-to-date patient medication (i.e., prescription drug) history on each of its customers to prevent a new prescription from causing an adverse reaction to a drug he or she is already taking. Here, we are concerned about real-time recognition – understanding at the point of service exactly who is the customer at the pharmacy counter.

For example, Mary Smith may have a prescription at the Albertsons store near her office, but need to refill it at another Albertsons – for example, the one near her residence. The pharmacist at the Albertsons near Mary's residence needs to know immediately what other medication Mary is taking. After all, there is at least one high-profile lawsuit per year against a pharmacy that results in at least a million dollar award. Given this concern, the return-on-investment (ROI) for the solution comes pretty rapidly.

In addition to health and safety issues, there are also issues involving the coverage of the prescription drug portion of the customer's health insurance. Albertsons has responded to both the safety and cost problems by deploying Initiate Identity Hub™ software to first identify and resolve duplication followed by implementing real-time access to patient profile data and prescription history throughout all its stores. This allows for a complete, real-time view of pharmacy-related information for its customers on

(1) the medications that are covered,
(2) the amount of the deductible, and
(3) the amount of the co-payment at the point of service to enable better drug utilization reviews and enhance patient safety.

RxHub provides a similar benefit at the point of service for healthcare providers accessing information from multiple pharmacy organizations. RxHub maintains records on 150 million patients/customers in the United States from a consortium of four pharmacy benefits managers (PBMs). Basically, everybody's benefits from pharmacy usage are encapsulated into a few different vendors.

RxHub takes the sum-total of those vendors and brings them together into a consortium. For example, if I have benefits through both my place of employment and my wife's place of employment, the physician can see all those in one place and use the best benefits available to me as a patient.

Even if my prescriptions have been filled across different PBMs, my doctor's office is able to view my complete prescription history as I come in. This is real-time recognition in its ultimate form. Because there is a consortium of PBMs, RxHub cannot access the data from the different members until it is asked for the prescription history of an individual patient. Then, in real time, RxHub identifies and links the records from the appropriate source files and consolidates the information on the patient for the doctor. RxHub is able to complete a search for an individual patient's prescription history in under $\frac{1}{4}$ second.

3.1.5.   Choice Hotels International

Choice Hotels International had built a data warehouse consisting entirely of its loyalty program users in order to analyze its best customers. Choice assumed that its loyalty program users were its best customers. Moreover, in the past, Choice could only uniquely identify customers who were in its loyalty program.

Then, Choice hired Initiate Systems, Inc. to analyze its data. Initiate Systems discovered that (1) only 10% of Choice's customers ever used a loyalty number and (2) only 30% of Choice's best customers (those who had at least two stays during a 3-month period) used a loyalty number. So, Choice's data warehouse only contained a small portion of its best customers.

Once Initiate Systems was able to implement software that uniquely identified Choice customers who had never used a unique identifier, Initiate Systems was able to give Choice a clearer picture of its true best customers. By using Initiate's solution, Initiate Identity Hub™ software, Choice now stores data on all of its customers in its data warehouse, not just the 10% who are members of its loyalty program.

For example, a customer might have 17 stays during a calendar year at 14 different Choice Hotels and never use a loyalty number. Because all the hotels are franchised, they all have their own information systems. So, the information sent to Choice's data center originates in different source systems and in different formats. Initiate Systems' software is able to integrate these data within the data warehouse by uniquely identifying Choice's customers across these disparate source systems.

3.2.   Data Quality Problems and their Consequences

Just as instructive as fruitful applications of high-quality databases – though having the opposite effect on the bottom line – are examples of real-world problems with bad data. We begin with one of the earliest published examples.

3.2.1. Indians and Teenage Widows

As reported by Coale and Stephan [1962], "[a]n examination of tables from the 1950 U.S. Census of Population and of the basic Persons punch card, shows that a few of the cards were punched one column to the right of the proper position in at least some columns." As a result, the "numbers reported in certain rare categories – very young widowers and divorces, and male Indians – . . .were greatly exaggerated."

Specifically, Coale and Stephan [1962] observed that

a shift of the punches intended for column 24 into column 25 would translate relationships to head of household (other than household head itself) into races other than white. Specifically, a white person whose relationship to the head was *child* would be coded as a male Indian, while a Negro child of the household head would be coded as a female Indian. If the white child were male, he would appear as an Indian in his teens; if female, as an Indian in his twenties. Since over 99% of "children" are under 50, and since the shift transfers [the] first digit of age in to the second digit of age, the erroneous Indians would be 10–14 if really male, and 20–24 if really female.

For example, in the Northeastern Census Region (of the United States), an area where the number of Indian residents is low, the number of male Indians reported by age group is shown in Table 3.1.

The number of male Indians shown in Table 3.1 appears to be monotonically declining by age group if we ignore the suspect entries for age groups 10–14 and 20–24. This leads us to suspect that the number of male Indians should be between (1) 668 and 757 for the 10–14 age group and (2) 596 and 668 for the 20–24 age group. This adds support to the conjecture that the number of male Indians in the 10–14 and 20–24 age groups is indeed too high.

For teenage males, there were too many widowers, as can be seen from the Table 3.2.

TABLE 3.1. Number of reported male Indians Northeastern US, 1950 census

Age (in years)								
Under 5	5–9	10–14	15–19	20–24	25–29	30–34	35–39	40–44
895	757	**1,379**	668	**1,297**	596	537	511	455

Source: Table 3, US Bureau of the Census [1953b]

TABLE 3.2. Number of (male) widowers reported in 1950 census

Age (in years)								
14	15	16	17	18	19	20	21	22
1,670	1,475	1,175	810	905	630	1,190	1,585	1,740

Source: Table 103, US Bureau of the Census [1953a]

In particular, it is not until age 22 that the number of reported widowers given in Table 3.2 exceeds those at age 14.

3.2.2.   IRS versus the Federal Mathematician

Over 30 years ago, a senior mathematician with a Federal government agency got a telephone call from the IRS.

IRS Agent: "You owe us $10,000 plus accrued interest in taxes for last year. You earned $36,000 for the year, but only had $1 withheld from your paycheck for Federal taxes."

Mathematician: "How could I work the entire year and only have $1 withheld? I do not have time to waste on this foolishness! Good-bye."

Question: What happened?

Answer: The Federal government agency in question had only allocated enough storage on its computer system to handle withholding amounts of $9,999.99 or less. The amount withheld was $10,001. The last $1 made the crucial difference!

3.2.3.   The Missing $1,000,000,000,000

A similar problem to the IRS case occurs in the Table 3.3.

Notice that the entry in the first column is understated by $1,000,000,000,000 because the computer software used to produce this table did not allow any entries over $999,999,999,999.99.

3.2.4.   Consultant and the Life Insurance Company

A life insurance company[4] hired a consultant to review the data quality of its automated policyholder records. The consultant filled out the necessary paperwork to purchase a small life insurance policy for himself. He was shocked when the company turned him down, apparently because the company classified him as "unemployed." The reason: he had omitted an "optional" daytime phone number and had only supplied the number for his cell-phone.

A more typical problem that this consultant reports concerns "reinstatements" from death. This frequently occurs on joint-life policies, such as family plans, where the policy remains in force after the first death claim. Shortly after the death of the primary insured, both the primary coverage status and the status of

TABLE 3.3. Status of insurance of a large insurance company

Insurance Written	Insurance Terminated	Insurance In-force
$456,911,111,110	$823,456,789,123	$633,454,321,987

---

[4] We thank Kevin Pledge, FSA, for providing the examples of this section

the entire policy are changed to "death." A month later, the surviving spouse's coverage status is changed to "primary" and the policy appears to have been "reinstated from death."

A final insurance example concerns a study of insurance claims on a large health insurance company. The company had an unusually high rate of hemorrhoid claims in its Northwest region. Further investigation revealed that the claim administration staff in this region did not think this code was used for anything and so used it to identify "difficult" customers. While this study may be apocryphal, our consultant friend reports many other cases that are logically equivalent, although not as amusing.

3.2.5. Fifty-two Pickup

One of the mortgage companies insuring its mortgages with the Federal Housing Administration (FHA) had 52 mortgages recorded on its internal computer system under the same FHA mortgage case number even though each mortgage is assigned its own distinct FHA mortgage case number. This prevented the mortgage company from notifying FHA when the individual mortgages prepaid. Needless to say, it took many hours for the mortgage company staff working together with FHA to correct these items on its computer system.

3.2.6. Where Did the Property Tax Payments Go?[5]

The Washington Post's February 6, 2005, edition reported that an unspecified computer error caused serious financial problems for a group of bank customers:

Three years ago in Montgomery County [Maryland], a mistake at Washington Mutual Mortgage Corp. resulted in tax payments not being correctly applied to 800 mortgages' property taxes. Most homeowners learned of the problem only when they received county notices saying that they were behind on their property taxes and that their homes might be sold off. The county later sent out letters of apology and assurances that no one's home was on the auction block.

3.2.7. The Risk of Massive ID Fraud

One day during May of 2004, Ryan Pirozzi of Edina, Minnesota, opened his mailbox and found more than a dozen bank statements inside. All were made

---

[5] This section and the next are based on an article by Griff Witte [2006] that appeared in the business section of the February 6, 2005, edition of *The Washington Post*.

out to his address. All contained sensitive financial information about various accounts. However, none of the accounts were his.

Because of a data entry error made by a clerk at the processing center of Wachovia Corp., a large bank headquartered in the Southeastern United States, over the course of at least 9 months, Pirozzi received the financial statements of 73 strangers all of whom had had escrow accounts with this bank. All of these people, like Pirozzi, bought real estate through the Walker Title and Escrow Company headquartered in Fairfax, Virginia. Their names, Social Security numbers, and bank account numbers constitute an identity thief's dream.

Then, during January 2005, Pirozzi began receiving completed 1099 tax forms belonging to many of these people. After inquiries from a reporter for *The Washington Post*, both Wachovia and the Walker Company began investigating the problem. This revealed that many people who purchased a condominium unit at the Broadway in Falls Church, Virginia were affected. These homebuyers were given a discount for using the developers' preferred choice, Walker, to close on the purchase of their condominium units. In order to secure a condominium unit in the new building, prospective homebuyers made deposits that were held in an escrow account at Wachovia.

The article in *The Washington Post* relates some comments of Beth Givens, director of the Privacy Rights Clearinghouse headquartered in San Diego.

Givens said that this case demonstrates that identity theft doesn't always stem from people being careless with their financial information; the institutions that people trust with that information can be just as negligent. Although the worst didn't happen here, information gleaned from misdirected mail can wind up on the black market, sold to the highest bidder.

There have been instances, Givens said, in which mail processing systems misfire and match each address with a name that's one off from the correct name. In those situations, she said, hundreds or even thousands of pieces of mail can go to the wrong address. But those kinds of mistakes are usually noticed and corrected quickly.

The Washington Post article also quoted Chris Hoofnagle, associate director of the Electronic Privacy Information Center, as saying,

It should be rather obvious when a bank sends 20 statements to the same address that there's a problem. But small errors can be magnified when you're dealing with very large institutions. This is not your neighborhood bank.

The article also reported that Mr Hoofnagle "said it would not be difficult for Wachovia to put safeguards in place to catch this kind of error before large numbers of statements get mailed to the wrong person." The article did not provide the specifics about such safeguards.

Finally, one day in January 2005, a strange thing happened. Mr Pirozzi went to his mailbox and discovered an envelope from Wachovia with his address and his name. It contained his completed 1099 tax form for 2004. "That" Pirozzi said "was the first piece of correspondence we received from [Wachovia] that was actually for us."

3.3.   How Many People Really Live to 100 and Beyond? Views from the United States, Canada, and the United Kingdom

Satchel Paige was a legendary baseball player. Part of his lore was that nobody ever knew his age. While this was amusing in his context, actuaries may be troubled because they can't determine the age of the elderly around the world. This makes it difficult to determine mortality rates. A related issue is whether the families of deceased annuitants are still receiving monthly annuity payments and, if so, how many?

On January 17–18, 2002, the Society of Actuaries hosted a symposium on *Living to 100 and Beyond: Survival at Advanced Ages*. Approximately 20 papers were presented at this conference. The researchers/presenters discussed the mortality experience of a number of countries in North America, Europe, and Asia. The papers by Kestenbaum and Ferguson [2002], Bourdeau and Desjardins [2002], and Gallop [2002] dealt with the mortality experience in the United States, Canada, and the United Kingdom, respectively. Many of the papers presented at this symposium dealt explicitly with data quality problems as did those presented at a follow-up symposium held during January, 2005. These studies illustrate the difficulty of obtaining an unambiguous, generally agreed-upon solution in the presence of messy age data.

3.3.1.   United States

Mortality of Extreme Aged in the United States in the 1990s, based on Improved Medicare Data, by Kestenbaum and Ferguson, describes how to improve the quality of information on old-age mortality in administrative databases by (1) careful selection of records, (2) combining sources (i.e., linking records from two or more databases[6]), and (3) obtaining supplemental information from other government agencies. According to Kingkade [2003], the authors' "meticulous" scheme "does not remove every indelicacy in the schedule of probabilities of dying, but it vastly improves the schedule's plausibility."

3.3.2.   Canada

In *Dealing with Problems in Data Quality for the Measurement of Mortality at Advanced Ages in Canada*, Robert Bourdeau and Bertrand Desjardins state (p. 13) that "after careful examination . . ., it can be said that the age at death for centenarians since 1985 in Quebec is accurate for people born in Quebec". On the other hand, the formal discussant, Kestenbaum [2003], is "suspicious" about "the accuracy of age at death" in the absence of "records at birth or shortly thereafter."

---

[6] Chapters 8 and 9 provide an extensive treatment of record linkage.

3.3.3.   United Kingdom

In *Mortality at Advanced Ages in the United Kingdom*, Gallop describes the information on old-age mortality of existing administrative databases, especially the
one maintained by the United Kingdom's Department of Work and Pensions. To
paraphrase the formal discussant, Kingkade [2003], the quality of this database
is highly suspect. The database indicates a number for centenarians that vastly
exceeds the number implied by a simple log of the Queen's messages formally
sent to subjects who attain their 100th birthday. The implication (unless one
challenges the authority of the British monarch) is that the Department's database
grossly overstates longevity.

3.4.   Disabled Airplane Pilots – A Successful Application of Record Linkage

The following example shows how record linkage techniques can be used to
detect fraud, waste, or abuse of Federal government programs.

A database consisting of records on 40,000 airplane pilots licensed by the US
Federal Aviation Administration (FAA) and residing in Northern California was
matched to a database consisting of individuals receiving disability payments
from the Social Security Administration. Forty pilots whose records turned up
on both databases were arrested. A prosecutor in the US Attorney's Office in
Fresno, California, stated, according to an Associated Press [2005] report, "there
was probably criminal wrongdoing." The pilots were "either lying to the FAA
or wrongfully receiving benefits."

"The pilots claimed to be medically fit to fly airplanes. However, they may
have been flying with debilitating illnesses that should have kept them grounded,
such as schizophrenia, bipolar disorder, drug and alcohol addiction and heart
conditions."

At least 12 of these individuals "had commercial or airline transport licenses."

"The FAA revoked 14 pilots' licenses." The "other pilots were found to be
lying about having illnesses to collect Social Security [disability] payments."

The quality of the linkage of the files was highly dependent on the quality
of the names and addresses of the licensed pilots within both of the files being
linked. The detection of the fraud was also dependent on the completeness
and accuracy of the information in a particular Social Security Administration
database.

3.5.   Completeness and Accuracy of a Billing Database: Why It Is Important to the Bottom Line

Six doctors have a group practice from which they do common billing and
tracking of expenses. One seemingly straightforward facet of data quality is the

main billing database that tracks (1) the days certain patients were treated, (2) the patients' insurance companies, (3) the dates that bills were sent to the patient, (4) the dates certain bills were paid, and (5) the entity that paid the bill (i.e., the health insurance company or the individual patient). Changes in the billing database are made as bills are paid. Each staff member in the practice who has access to the database (via software) has a login identifier that allows tracking of the changes that the individual made in the database. The medical assistants and the doctors are all given training in the software. One doctor acts as the quality monitor.

In reviewing the data in preparation for an ending fiscal year, the doctors realize that their practice is not receiving a certain proportion of the billing income. Because of the design of the database, they are able to determine that one doctor and one medical assistant are making errors in the database. The largest error is the failure to enter certain paid bills in the database. Another error (or at least omission) is failure to follow-up on some of the bills to assure that they are paid in a timely manner. After some retraining, the quality monitor doctor determines that the patient-billing portion of the database is now accurate and current.

After correcting the patient-billing portion, the doctors determine that their net income (gross income minus expenses) is too low. They determine that certain expenses for supplies are not accurate. In particular, they deduce that (1) they neglected to enter a few of their expenses into the database (erroneously increasing their net income), (2) they were erroneously double-billed for some of their supplies, and (3) a 20% quantity discount for certain supplies was erroneously not given to them.

3.6.   Where Are We Now?

These examples are not a random sample of our experiences. We chose them to illustrate what can go right and, alas, what can go wrong. One of the reasons we wrote this book is that we believe good experiences are too infrequent and bad ones are too common.

Where are you? Our guess is that since you are reading this book you may share our concerns. So, we hope that you can use the ideas in this book to make more of your personal experiences good ones.

4
Properties of Data Quality and Metrics for Measuring It

Metrics for measuring data quality (or lack of it) are valuable tools in giving us some quantitative objective measure of the problems we may be dealing with. In this chapter, we first discuss a few key properties of high-quality databases/lists. This is followed by a number of typical examples in which lists might be merged. Finally, we present some additional metrics for use in assessing the quality of lists produced by merging two or more lists.

Although quantification and the use of appropriate metrics are needed for the quality process, most current quantification approaches are created in an ad hoc fashion that is specific to a given database and its use. If there are several uses, then a number of use-specific quantifications are often created. For example, if a sampling procedure determines that certain proportions of current customer addresses are out of date or some telephone numbers are incorrect, then a straightforward effort may be needed to obtain more current, correct information. A follow-up sample may then be needed to determine if further corrections are needed (i.e., if the database still lacks quality in some respect). If no further corrections are needed, then the database may be assumed to have an acceptable quality for a particular use.

4.1. Desirable Properties of Databases/Lists

Ideally, we would like to be able to estimate the number of duplicate records as well as the number of erroneous data items within a database/list. We would like every database/list to be complete, have few, if any, duplicate records, and have no errors on the components of its data records. If a corporation's mailing list of business or retail customers is incomplete, this could lead to lost business opportunities. If the list consists of subscribers to a magazine, the omission of subscribers from the magazine's mailing list could lead to extra administrative expense as the subscribers call the company to obtain back issues of magazines and get added to the mailing list.

Duplicate records on corporate databases can lead to extra printing and postage costs in the case of a mailing list of retail customers or double billing of customers on a corporate billing system. Duplicate records on the database of a mutual insurance company could lead to duplicate payment of dividends to policyholders. It is important as well to eliminate duplicate names from a corporation's mailing list of current stockholders in order to improve the corporate image.

If a list is used as a sampling frame for a sample survey, an incomplete list might lead to under-estimates of the quantities of interest while duplicate records could lead to over-estimates of such items. Duplicate records could also lead to duplicate mailings of the survey questionnaire. Examples of sampling frames considered later in this text include the following:

- A list of all of the sellers of petroleum products in the United States – see Section 12.3.
- A list of all of the agribusinesses (i.e., farms) in the United States – see Section 16.2.
- A list of all of the residents of Canada – see Section 16.1.

Examples of errors on individual data elements include the wrong name, wrong address, wrong phone number, or wrong status code.

Three simple metrics can be usefully employed for assessing such problems.

The first metric we consider is the *completeness* of the database/list. What proportion of the desired entities is actually on the database/list?

The second metric is the *proportion of duplicates* on the database/list. What proportion of the records on the database/list are duplicate records?

The third metric is the proportion of each data element that is missing.

Example 4.1: Householding

The husband, wife, and their three minor children living at 123 Elm Street might all be IBM stockholders. Instead of sending five copies (one to each member of the family) of its annual report each year together with accompanying proxy material for its annual stockholders' meeting, IBM might be able to send just a single copy of its annual report and to consolidate the required proxy material into a single envelope. Such consolidation is called "householding."

A public corporation (Yum! Brands – the owner of such brands as KFC, Taco Bell, and Pizza Hut) recently mailed out the following letter to its shareholders.

> Dear Yum! Shareholder:
>
> The Securities and Exchange Commission rules allow us to send a single copy of our annual reports, proxy statements, prospectuses and other disclosure documents to two or more shareholders residing at the same address. We believe this *householding* rule will provide greater convenience for our shareholders as well as cost savings for us by reducing the number of duplicate documents that are sent to your home.
> Thank you

Other examples of mailing lists include those of professional organizations (e.g., the American Statistical Association or the Society of Actuaries), affinity groups (e.g., the American Contract Bridge League or the US Tennis Association), or alumni groups of Colleges, Universities or Secondary Schools.

4.2. Examples of Merging Two or More Lists and the Issues that May Arise

In this section, we consider several examples in which two or more lists are combined into a single list. Our goal in each case is a single, composite list that is complete and has (essentially) no duplicate entries.

Example 4.2: Standardization

A survey organization wishes to survey individuals in a given county about retirement and related issues. Because publicly available voter registration lists contain name, address, and date of birth, the organization seeks to obtain all of the voter registration lists needed to cover the entire county. These lists will then be merged to create a list or sampling frame of all county residents between ages 50 and 75. The survey organization plans to download these voter registration lists from public sites on the Internet.

Several issues need to be addressed in such an effort. The first is whether the survey organization will be able to identify all requisite voter registration lists for the county. If certain needed lists are not located, then some areas of the county may not be surveyed. This can adversely affect the survey, especially if certain low-income or high-income areas are not on the list that serves as the sampling frame.

The second issue is whether the voter registration lists contain all individuals between ages 50 and 75 or a representative subset of such individuals. An individual who moves to a different part of the same county may be on two or more distinct voter registration lists. The survey organization needs to assure that no voter is on a list more than once. Moreover, the organization needs to find the most recent address of every individual on the merged list. The organization plans to use the status codes associated with the address field to eliminate prior addresses (for those who relocate) within the county. The deletion of outdated addresses may be difficult if the status codes and address formats differ substantially across the lists. Here, suitable metadata describing the specific format of the address, status code, name, and date-of-birth fields on each of the lists may be crucial. So, it may be necessary to recode certain data elements in a common formatting scheme in order to facilitate the record linkage process. For example, if one list has a date of birth as "January 12, 1964" and another as "120164" (where the second date of birth has the form DDMMYY), then some recoding needs to be done to make these data elements compatible. This type of recoding is known as *standardization*. Standardization and a companion technique known as *parsing* are both discussed in-depth in Chapter 10.

Example 4.3: Combining Mailing Lists

A mail-order company maintains a list of retail customers from its traditional mail-order business that it wishes to (seamlessly) combine with a list of customers compiled from its new Internet business operation. Additionally, the company wishes to combine its customer list with lists from external sources to determine subsets for sending targeted advertising material. We assume that the traditional mail-order part of the business has good computer files that are complete and unduplicated. At a minimum, the company wants to carry the same name, address, dates of purchases, items ordered, and account numbers for both the traditional mail-order and Internet business. For orders that are mailed in, the company has good procedures for keying name, address, account number, and credit card information. For orders that are called in, the company has good procedures for assuring that its telephone agents key in the information accurately. The Internet site requests the same information that includes whether someone is a prior customer (a check box) and the customer's account number. If most of the information in all of the components of the database is correct, then the company can effectively combine and use the information.

But there are some instances in which quality can deteriorate. For example, the mail-order portion of the mailing has a listing of "Susan K. Smith" at "123 Main St." This listing was obtained from a mailing list purchased from another company. The Internet portion of the list may have a customer listed as "Karen Smith" because the individual prefers to use the name "Karen." She is listed at a current address of "678 Maple Ave" because she has recently moved. In such situations, customers may be listed multiple times on the company's customer list. If a customer's account number or a telephone number is available, then the mail-order company may be able to identify and delete some of the duplicate customer entries.

There are several issues. Can the company track all of its customers or its most recent customers (of the last 18 months)? Does the company need a fully unduplicated list of customers?

Example 4.4: Combining Lists and Associated Data Fields

Two companies merge and wish to consolidate their lists of business customers into a single list. They also want to combine data fields associated with the lists. If the name and business address of a customer on the first company's list is given as "John K. Smith and Company, PO Box 5467" and the same customer in the second company's list is given as "J K S, Inc., 123 Main St" where "123 Main St" is the address of the company's accountant, then it will be very difficult to combine the customer lists. A partial solution might be to carry several addresses for each customer together with the date of the most recent transaction associated with that address. The addresses and dates might need careful review to assure that the current best address for contacting the customer is in the main location. An analogous situation occurs when an organization that conducts sample surveys wishes to either (1) consolidate several of its list frames or (2) merge one of its list frames with an external list.

Example 4.5: Combining Lists of Products for Sale to the Public

Two companies merge and wish to combine the list of the products they sell into a single list. If one product is listed as an "adaptable screwdriver" and another product is listed as a "six-purpose screwdriver," how will the merged company know whether the two products are identical (possibly from the same manufacturer) or completely different products?

4.3.   Metrics Used when Merging Lists

In this section, we describe metrics that can be used with various types of databases. It is important to realize that metrics that are appropriate for one database may be inappropriate for a similar database with a different error structure. However, the analyst can readily construct new metrics tailored to her specific application. The construction of such metrics is a fertile area for future research.

4.3.1.   Notation

We let two lists be denoted by A and B. If a record $a \in A$ and a record $b \in B$, then we say the pair of records (a, b) is a "true" matching pair if both records represent the same entity. For example, both records might represent the corporate headquarters of a major computer software company. We let

$$M = \{(a, b) : a \in A, \ b \in B, (a, b) \text{ is a true matching pair}\}$$

and

$$U = \{(a, b) : a \in A, \ b \in B, (a, b) \text{ is not a matching pair}\}.$$

So, the sets M and U form a partition of the cross-product space $A \times B$. Hence, we can write $A \times B = M \cup U$.

We might want to compare two such lists to identify possible matching pairs of records. Here, we let

$$\widetilde{M} = \{(a, b) : a \in A, \ b \in B, (a, b) \text{ is designated as a matching pair}\}$$

and

$$\widetilde{U} = \{(a, b) : a \in A, \ b \in B, (a, b) \text{ is not designated as a matching pair}\}.$$

Similarly, the sets $\widetilde{M}$ and $\widetilde{U}$ form a partition of the cross-product space $A \times B$. Hence, we can write $A \times B = M \cup U = \widetilde{M} \cup \widetilde{U}$. Of course, in practice it is highly unlikely that we will correctly identify all of the matches in M. So, we will usually have $\widetilde{M} \neq M$ and $\widetilde{U} \neq U$.

4.3.2. The Metrics

The *false match rate* is the proportion of actual non-matches designated as matches:

$$P[(a, b) \in \widetilde{M} | (a, b) \in U].$$

The *false non-match rate* is the proportion of actual matches that are designated as non-matches:

$$P[(a, b) \in \widetilde{U} | (a, b) \in M].$$

The *precision* is the proportion of designated matches that are actual matches:

$$P[(a, b) \in M | (a, b) \in \widetilde{M}].$$

We note that

$$P[(a, b) \in M | (a, b) \in \widetilde{M}] + P[(a, b) \in U | (a, b) \in \widetilde{M}] = 1$$

where by Bayes' Theorem[1] we can obtain

$$P[(a, b) \in U | (a, b) \in \widetilde{M}] = \frac{P[(a, b) \in \widetilde{M}(a, b) \in U] \cdot P[(a, b) \in U]}{P[(a, b) \in \widetilde{M}]}.$$

The *recall rate* is the proportion of actual matches that are designated matches:

$$P[(a, b) \in \widetilde{M} | (a, b) \in M].$$

We note that the sum of the false non-match rate and the recall rate is one:

$$P[(a, b) \in \widetilde{U} | (a, b) \in M] + P[(a, b) \in \widetilde{M} | (a, b) \in M] = 1.$$

The probability function, $P[\cdot]$, that we use here is a relative frequency function in which all events are assumed to be equally likely to occur.

Example 4.5: Tradeoffs May Need to Be Considered

If the false non-match rate is too high (i.e., if $\widetilde{U}$ is too large) for the merged lists of Examples 4.3 and 4.4, then the merged lists will have some customers listed twice. So, the total number of customers could be over-stated with a consequence being that future sales estimates are too high. Nevertheless, a company might prefer to have duplicates on its mailing list rather than to omit customers. (In other words, the company would like the false non-match rate to be higher than the false match rate.) The company may feel that erroneous forecasts and (possible) increased printing and postage expense on mailed advertising material is preferable to actual lost sales. The key issue here is making these trade-offs as explicit and numerically exact as possible.

---

[1] See Section 6.2 for a description of Bayes' Theorem.

4.4.   Where Are We Now?

In this chapter, we have presented a number of metrics that can be used to give us quantitative, objective measures of the quality of our data. In particular, we have presented a number of metrics that can be used to quantify the effectiveness of record linkages.

5
Basic Data Quality Tools

In this chapter, we describe a number of basic data editing techniques. We begin by defining a *data element*. We then discuss a number of different types of *deterministic tests* applied to these data elements. We conclude with a brief discussion of both *probabilistic tests* and *exploratory data analysis* techniques.

Many data systems employ data editing techniques. For simple data systems, the editing procedures may consist of a few basic checks and tests. As such systems increase in both size and complexity, however, the number of such tests usually increases as well. Large complex databases, containing many different types of data elements, might require several hundred interrelated tests.

When dealing with large databases, the number of tests often grows rapidly. When a large number of tests is involved, they need to be organized into a dictionary. The effort compiling such a dictionary can be minimized if the data editing is organized from the beginning in an orderly fashion. Even in large systems this is often not done. We consider the organizational problem in more detail in Chapter 7, our chapter on editing and imputation. The interested reader could also refer to Chapter 2 of Naus [1975].

5.1.   Data Elements

The first step in the construction of an editing/imputation system is the creation of a list of the data elements contained within the data system. A *data element* or a *data field* is defined as an aspect of an individual or object that can take on varying values among individuals. Every piece of information in the data system can be viewed as a measurement on a data element.

In most data systems, there are several distinct types of data elements. Some data elements can only assume a discrete set of values (i.e., a countable number of values); other data elements can be viewed as having an underlying continuous set of possible values taken from one or more continuous intervals of values. Sometimes, the values of the data elements are numeric – that is, quantities or magnitudes; other times, these values are qualities. When organizing the edits, it is useful to distinguish between qualitative data elements and quantitative data elements.

The different types of data elements can be illustrated by a simple example. In Table 5.1, we consider a few data elements on a small number of case records in a database of mortgages.

TABLE 5.1. Extract of database of FNMA/FHLMC[1] conforming mortgages

Name of primary borrower	Year of mortgage origination	Month of mortgage origination	Gender of primary borrower	Term of mortgage (in months)	Annual contract interest rate (in %)	Original mortgage amount (in $)
Jane Smith	1999	08	F	360	6.5	250, 000.00
Robert Brown	2001	06	M	360	6.25	275, 000.00
Harry Turner	2000	03	M	240	5.875	230, 000.00
Mary Sunshine	2003	02	F	300	6.25	300, 000.00
Susan Stone	2002	11	F	360	6	295, 000.00

Some data elements can only take on a finite number of (admissible) values or codes. For example, the data element on the "gender of the primary borrower" can only be one of two values – "M" or "F". Some data elements such as month of mortgage origination have a natural set of values – the month is coded to be a two-digit number between 01 and 12. The quantitative data element "annual contract interest rate" can, in theory, assume positive values in a continuous range. However, in practice, it rarely exceeds five decimal places – for example, a typical value in calendar year 2005 was .05875 or 5.875%, whereas in calendar year 1981 a more typical value was .15875 or 15.875%. The quantitative data element "original mortgage amount" can only take on a finite number of values because, by statute, there is a maximum amount of a conforming mortgage and the amount must be in dollars and cents. During 2006, the maximum amount of a conforming single-family loan on a home located in one of the 48 contiguous States was $417,000.

5.2.   Requirements Document

Much of the idea of data quality begins with the creation of a conceptual framework that answers the following questions:

- Why should the database be established?
- How are the data to be collected?
- How do analysts plan to use the data?
- What database-design issues may affect these data uses?

As a possible first step, the potential users may simply make a list of requirements for the data and associated system. This list is created to help ensure that the system designers will meet the needs of potential users. The final list may evolve into a requirements document that provides an overview of the system design.

---

[1] FNMA = Federal National Mortgage Association and FHLMC = Federal Home Loan Mortgage corporation.

The requirements document lists some (or most) of the components needed for collecting data and entering the data into a database. It contains sufficient detail so that the software engineers, database designers, and others will know the requirements of a system. By database, we mean the databases of computer science, SAS databases, or sets of sequential (flat) files. In order to produce a requirements document, potential users will have to meet with software engineers, database designers, and other individuals who can provide specific advice.

This process may be expedited by employing an experienced facilitator (possibly someone who has already worked on similar projects) to improve the communications among these individuals. The facilitator can ensure that all critical items are on the list and individual stakeholders are communicating in a fashion that allows understanding. For instance, an experienced systems programmer may be able to explain how various characteristics in the database can help customer-relation representatives and data analysts who may be interested in marketing. An experienced customer-relation representative may be able to identify certain data fields that must be readily available during interactions with customers. The marketing representative may want various pop-up screens that allow the customer-relation representatives to see additional products.

In particular, the requirements document may include the initial steps in developing the codebook portion of the "metadata" for the system.[2] Metadata contain descriptions of the fields in the database and their permissible values, as well as how they are created and limitations on their use, if known.

For instance, the metadata for the concept "Name" may say that a name is stored as four distinct components: last name, middle initial, first name, and title. The metadata for "Address" may specify that the US Postal Zip Code must be checked against a suitable database to ensure the accuracy of the Zip Code and (possibly) the existence of the address in the form of house number and street name. Although primarily qualitative, the requirements document and metadata often are the starting point for assuring minimally acceptable quality in a system.

5.3. A Dictionary of Tests

In the explanation of Table 5.1, all the data elements have explicit or implicit bounds on the set of admissible values. Such bounds define regions of acceptance or rejection that can be incorporated into deterministic tests on the individual data elements.

In constructing a dictionary of tests, the analyst may consult a number of subject matter experts and may investigate various other sources of information. The analyst typically compiles a list of acceptable and unacceptable values of individual data elements as well as a list of permissible combinations of values of data elements. For example, a test may be defined and then implemented by determining whether the observed combination of data element values on

---

[2] For more information on metadata, see Dippo and Sundgren [2000] or Scheuren [2005a and 2005b].

an individual data record is either (1) within the acceptable region of values or (2) in the rejection (unacceptable) region of values. If the data are within the unacceptable region, we say that the test "rejects" the data. In such circumstances, we might simply "flag" either the problematic data element or the entire data record; alternatively, we may not enter a data record onto the data system until its data are made acceptable.

5.4. Deterministic Tests

In this section we describe a number of types of deterministic tests to help us edit our data. We include a number of examples of each such test.[3]

5.4.1. Range Test

The simplest test is one in which the regions of acceptance and rejection of the test involve the values of only a single data element. For example, the test accepts the datum for the "Month" data element if the value coded for "Month" is an element of the set $\{01, 02, 03, 04, 05, 06, 07, 08, 09, 10, 11, 12\}$; otherwise, the test will reject the datum. Such a test, which checks whether the value of a single data element is within a range of values, is called a *range test*.

If the gender code of an employee database is given by

$$0 = \text{``missing''}, \ 1 = \text{``male''}, \text{ and } 2 = \text{``female''},$$

then values outside of the set $\{0, 1, 2\}$ are erroneous.

If the salary range of an employee in a specific job category is known to be within an interval [L, U], then any salary outside that interval is erroneous.

5.4.2. If-Then Test

The next test we consider is of the following type: If data element X assumes a value of x, then data element Y must assume one of the values in the set $\{y_1, y_2, \ldots, y_n\}$. For example, if the "type of construction" of a house is "new," then the age of the house can not be a value that is greater than "1" year. If the age of the house takes on a value that is greater than "1" year, then we must reject the pair of data element values. This is an example of an *if-then test*.

A few other examples of this type of test, typical of data encountered in a census of population or other demographic survey, are as follows:

If the relationship of one member of the household to the head of the household is given as "daughter", then the gender of that individual must, of course, be "female".

If the age of an individual is less than 15, then the marital status should be "single". In many western European countries, Canada and the United States, an edit-correction may always require that the value of the marital status field for an individual under age 15 be "single". Although this requirement may create an error in a small number of records,

---

[3] These tests were discussed previously in Naus [1975]. His classification scheme gives us a useful framework for categorizing such tests.

the aggregate effect of this is almost always a substantial improvement in the quality of the database.

If the ages of the parents (head of household and spouse) are each approximately twenty five years, then the age of each of their children should be considerably less (say, at least 15 years less).

If the age of the wife is more than twenty years greater than the age of the husband, then check both ages.

In repeated applications, the performance of edits themselves should be measured and evaluated. In many situations, if we have extensive experience analyzing similar data sources, we might decide to exclude certain edits on errors that occur rarely – for example, one time in 100,000. The example above comparing the age of the wife to the age of the husband might be an example of this.

5.4.3. Ratio Control Test

We next describe a class of procedures that considers combinations of quantitative data elements. In particular, these tests determine whether such elements satisfy certain arithmetic constraints. The first of these is known as *the ratio control test*.

We first illustrate the application of the ratio control test with economic data as follows. Suppose one data element is the number of bushels in a shipment of wheat and the other data element is the value (or wholesale cost) of the shipment in dollars. Then, based on our knowledge of the market for wheat, we may be able to put bounds U and L around the ratio

$$\frac{\text{value in dollars of shipment}}{\text{number of bushels of wheat in shipment}}$$

as follows:

$$L < \frac{\text{value in dollars of shipment}}{\text{number of bushels of wheat in shipment}} < U.$$

This ratio represents the value (or wholesale price) per bushel of wheat. The scalars L and U are, respectively, the lower and upper bounds of the acceptance region of this ratio control test.

Another such example concerns the average salary of employees of a company. If we know both (1) the number of employees of a company and (2) its annual payroll, we can obtain an estimate of the average annual salary of the employees of that company by dividing (2) by (1). So, if we have such data for a large number of companies in an industry, then we can determine upper and lower bounds for an acceptance region on the average annual salary of employees of companies in this industry.

To be more specific, suppose that (1) we have gathered data on "n" companies in an industry, (2) P_i is the total annual payroll of the ith company in the industry, and (3) T_i is the average number of employees of the ith company during the year, where $i = 1, 2, \ldots, n$. Here again we expect

$$L < \frac{P_i}{T_i} < U \quad \text{for each } i = 1, 2, \ldots, n$$

where L and U are respectively the lower and upper bounds. The basic concept is that within a particular industry, the ratio P_i/T_i (representing the average annual salary of employees at the ith company) should be in a limited range. In practice, reporting errors and/or transcription/data entry errors may result in erroneous values of P_i or T_i that can have a large effect on the corresponding ratio. This is more likely to occur during the first year of a new sample survey. If we have a large set of values of the ratios P_i/T_i from surveys conducted during prior years, then we can make appropriate adjustments to these ratios to account for industry trends over time.

The bounds L and U can be established separately for each industry of interest. Within an industry, the bounds may be established by examining targeted subsets of companies such as the largest and smallest ones because the larger companies may have different characteristics (in terms of edits) than the smaller ones. For ongoing surveys, the bounds can also be established using survey data from the current time period.

5.4.4. Zero Control Test

Another relationship test – *the zero control test* – using several data elements is sometimes useful for control purposes. This method has its roots in accounting work in which both component elements and their sums are recorded separately in a data system. For example, a navy ship may report that its crew consists of 20 enlisted personnel and 3 officers. If a subsequent report gives a total of 25 crew-members, then the data fail this test. In this example, certain items need to add to a particular total.

Another example of this type of test is "The Missing $1,000,000,000,000" example considered earlier in Section 3.2. In that example, because the amount of insurance written was required to equal the sum of the current amount of insurance in force plus the amount of insurance previously terminated, we discovered that the amount of insurance written had been understated by one trillion dollars.

A third example concerns the monthly payment on a single-family mortgage. The monthly mortgage payment must be equal to the sum of (1) the interest payment, (2) the principal payment, and (3) the amount of money put into escrow for future payment of property taxes and/or homeowners insurance.

Sometimes, this test is relaxed slightly so that the sum of the reported component elements is only required to be close (i.e., within a tolerance level) to the reported/required sum. For example, in a recent survey (see Mustaq and Scheuren [2005]) assessing the cost of collecting State sales tax, retail businesses were asked to report their percentage of sales made via (1) their retail stores, (2) their catalogue, and (3) the Internet. The three component elements were deemed to be acceptable if their sum was between 90 and 110%; otherwise, all three component elements were treated as missing.

Balancing is another name for a scheme in which items must add to a total. The associated equation is sometimes called a *balance equation*. Whereas the examples above deal only with one-dimensional situations, this test could be extended to situations involving sums along two or more dimensions, for example a multi-dimensional contingency table. *Single-level balancing* is used to mean that no data element (field) of a database may occur in more than one balance equation of the editing scheme used for that database.

5.4.5. Other Internal Consistency Tests

The "if-then test" and "zero control test" considered above are examples of internal consistency tests – that is, they check to see if the values of the data elements of a given record within the database are consistent. Besides the specific examples considered above, many other internal consistency tests readily come to mind. For example, age and date of birth should be consistent, as should the Zip Code of the property address of a house and the State in which it is located.

Additionally, given name and gender should be consistent. For example, a likely problem in an employee database is a record in which the employee's name is listed as "Susan Smith" (i.e., the employee's given name is "Susan") and the gender is equal to "1" for "male." Moreover, if a record in that database has the number of years worked greater than or slightly less than the age of the employee, then either the age element or the number of years worked element is likely in error.

In health studies, the gender and medical procedure and/or disease fields should be consistent in the sense that some medical procedures and/or diseases are confined to one gender – for example, pregnancy or prostate-related issues.

A company that sells homeowners insurance may issue a policy number in which the first four digits of the number represent the State in which the house is located. So, for individual policies, the policy number and the State should be consistent in this respect. If the policy number is entered into the insurance company's data system incorrectly, then problems may well arise later on. For example, if a claim is filed under that policy and the policy cannot be located on the data system, then it is possible that a new record could be created on the data system with the correct policy number. This would lead to duplicate records on the insurance company's data system.

Another type of problem that might arise is if a mail-order company recorded someone's name twice under slightly different variations – for example, Thomas Jones and Tom N. Jones. This could lead to two separate records on the company's mailing list and thereby cost the company extra printing and mailing expense, and perhaps even some ill-will.

In the last two examples, duplicate records have polluted a corporate data system. A technique that can be employed to clean up (or *de-dupe*) data systems with such problems is called *record linkage*. This is the topic of Chapter 8. Chapter 8 also describes other uses of record linkage techniques – techniques that involve linking records from two or more data systems.

5.5. Probabilistic Tests

There is an extensive literature on probabilistic tests in general and tests for detecting outliers – atypical, infrequent observations – in particular. In fact, there is at least one text – Barnett and Lewis [1994] – whose focus is on outliers in statistical data. The intuition is that errors in data are likely to appear as outliers. If we plot the observed values of a univariate statistic, then the outliers are likely to appear in either the lower- or upper-tail of the distribution. Similarly, outliers of bivariate statistics can be identified using fairly simple graphical techniques.

For example, we could employ a ratio edit on the ratio of an individual's total monthly income to the total number of hours worked that month within a given job category. The resulting hourly wage rate must not be too high or too low. In some situations, clerical staff may review these wage rates and follow up manually to correct erroneous values. In other situations, the automated editing procedures of Chapter 7 may be used to change all of the values that fail edits.

Some recent works on probabilistic tests for outliers include Rousseeuw and Leroy [2003, especially Chapter 6] as well as Dasu and Johnson [2003, especially Sections 5.2.3 and 5.2.5]. Except for the brief discussion of exploratory data analysis in Section 5.6, we do not discuss the important topic of probabilistic tests further because it is so well covered in the published literature.

5.6. Exploratory Data Analysis Techniques

John Tukey of Princeton University and Bell Laboratories was the leading proponent of an area of statistics called *Exploratory Data Analysis* (EDA). EDA is an approach/philosophy toward data analysis that employs various techniques – primarily graphical – to, among other things, detect outliers, extreme values, and other anomalies in a set of data. Besides Tukey [1977], other references on EDA include Mosteller and Tukey [1977], Velleman and Hoaglin [1981], and Cleveland [1994].

The researchers at NIST [2005] contrast classical statistical analysis and EDA as follows.

For classical analysis, the sequence of steps in the process is

$$\text{Problem} \Rightarrow \text{Data} \Rightarrow \text{Model} \Rightarrow \text{Analysis} \Rightarrow \text{Conclusions}$$

For EDA, the sequence of steps in the process is

$$\text{Problem} \Rightarrow \text{Data} \Rightarrow \text{Analysis} \Rightarrow \text{Model} \Rightarrow \text{Conclusions}$$

Some practitioners might argue that this is only a philosophical difference, and that it would not make much difference in the solution of practical problems. But consider the conversation below between two modelers.

Modeler I: I just discovered that the data system we have been working on for the last five years has major data quality problems.

Modeler II: That is why I treat data systems the same way I do sausage – I do not want to know what is inside either one.

Modeler I: Ouch!! That is why I am a vegetarian!

We need to make a final observation here. Both sequences of steps listed above usually result in an iterative process. For example, for the EDA sequence of steps, after we formulate our initial set of conclusions, we may go back and do additional analyses, construct new models, and formulate new conclusions. We could repeat this process a number of times until we decide we have a reasonable solution to the problem at hand. We might even decide to collect additional data.

The box and whiskers plots shown below are typical of graphs used in exploratory data analyses. As you can see, these are excellent tools for comparing empirical probability distributions.

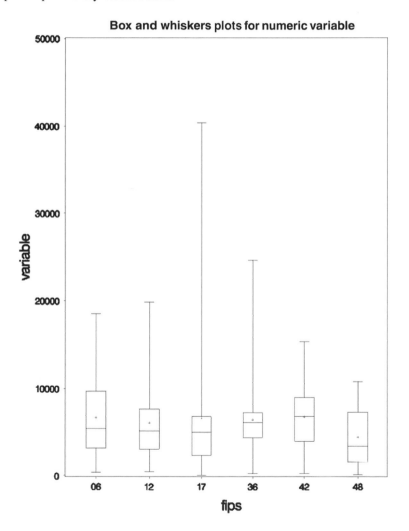

5.7.   Minimizing Processing Errors[4]

Deming [2006] has suggestions for reducing processing errors. Because he formulated the argument so well, rather than attempting our own version, we extract his words here.

A review of the codes assigned on a schedule is oftentimes not a matter of correcting wrong codes, but merely a matter of honest differences of opinion between coder and reviewer. Two coders will often find themselves in disagreement on the correct codes to assign to a response. Two coders working on the same set of schedules are going to turn out two different sets of results; likewise two different sections of coders working on the same set of schedules are going to turn out two different sets of results. *A fortiori*, two sections of coders working under slightly different instructions will show still greater differences, even though the two sets of instructions supposedly say the same thing in different words. The two sets of results may however produce distributions so nearly alike that in most problems they would lead to the same action, and that is what counts. Research needs to be carried out to show the extent of the differences to be expected from various shades of wording of instructions for coding, editing, and fieldwork. The conclusion seems inevitable that unless it is merely a matter of transcription (such as 1 for male and 2 for female) it is impossible to define a perfect job of coding except in terms of the distributions produced because there is no way of determining whether the individual codes have been assigned correctly. One can only say that two different sets of instructions or two different sets of coders produced substantially the same set of distributions. In view of this fact it seems to follow that when the work of the coder or editor or punch operator is uniformly good enough so that his errors are relatively insignificant compared with the other errors (such as variability of response) it is only necessary to perform enough review of his work (preferably by sampling methods) to be assured of continuity of control. . . .Workers who cannot qualify for sample review should be transferred.

   Machine and tally errors are often supposed to be negligible or non-existent but the actual situation is otherwise. These errors can be held to a reasonable minimum, however, by machine controls and other checks, especially with a force of workers in which there are a few key people with seasoned experience.

   A sample study or other partial coverage possesses a distinct advantage in the processing for the same reason it does in the interview, viz., the smaller force required to do the work, and the consequent better control that is possible.

5.8.   Practical Tips

We strongly recommend that data analysts organize their tests systematically and proceed in a way that (1) allows them to readily correct obvious errors that are easy to fix but (2) leaves more complicated situations for the more sophisticated methods of Chapter 7, our chapter on editing and imputation.

---

[4] Deming [2006] is a condensed version of Deming [1944] that Scheuren edited.

We also recommend that the analysts preserve the original dataset and also save intermediate versions of the dataset periodically. This can save countless hours of rework.

We suggest that the analyst keep tallies[5] or counts of "errors." Not all errors need to be corrected – only those that keep the dataset from being fit for an important use.

Many managers of large data systems give in to the temptation to over-edit. Over-editing can be very expensive in time and money. It may not even lead to a measurably improved final dataset. The tragedy here is the lost opportunity cost of using staff expertise to repair the final product instead of improving the system.

Consistency checks can be highly effective in detecting data quality problems within data systems. Therefore, it is important to design data systems[6] in such a fashion that such tests can be performed. One way that some survey practitioners accomplish this is by building in redundancy. One example is to ask the interviewee to provide both his/her age and date of birth. Another example is to record the street address, city, state, and Zip Code of a property address.

There are two challenges for practitioners: (1) to create innovative redundancy schemes – preferably, highly automated ones – for improving the quality of the data systems under their purview and (2) to use such systems frugally, and only when a virtually error free result is required. Otherwise, the expenses of such a system may be onerous.

Example 5.2: Using Multiple Addresses to Eliminate Duplicate Record Entries

A department store wants its database to carry the current and two prior addresses (with appropriate status flags and input dates) associated with customers. The multiple addresses make it easier to verify information for a given customer and to unduplicate accounts with commonly occurring names such as "John Smith." The users of the system may make explicit requests about the form of the names and addresses of customers as well as how, specifically, they want the names and addresses cleaned to a sufficient level of accuracy and currency.

If even a moderate proportion of the "latest" names, addresses, and telephone numbers is out of date or erroneous, then it may be difficult for customer service representatives to contact the customers or to use the list for the mailing of catalogs. If the customer list also contains, or can easily be associated with, certain marketing information, then the mailing of specialty catalogs to targeted customers may be more cost effective.

---

[5] In a survey sampling setting these counts should be both weighted and unweighted. Irrespective of the setting, the analyst needs to focus attention on assessing whether the repeated edits (iterations) continue to improve the data in material ways. Usually, after the first few major errors are corrected, the law of diminishing returns sets in rapidly.

[6] In this section, we make a few brief comments about the design of data systems. While this is an important topic, we feel that further comments are outside the scope of our work.

5.9.   Where Are We Now?

Because this chapter is the essence of our discussion on data quality, let us review what we have discussed. This chapter began with a formal definition of a data element and then discussed the creation of a conceptual framework for the database/list to be constructed. We devoted the bulk of the chapter to deterministic tests. These included range tests, if-then tests, ratio control tests, zero control tests, and internal consistency tests. In lieu of discussing probabilistic tests, we referred the reader to the abundant literature in this area except for a brief discussion of exploratory data analysis techniques. We also included a discussion of ways to build in redundancy in order to enhance our ability to use automated schemes to minimize data errors within databases/lists. We concluded the chapter with some material we extracted from Deming [2006] on processing errors.

Part 2
Specialized Tools
for Database Improvement

6
Mathematical Preliminaries for Specialized Data Quality Techniques

In Part One of this book we defined data quality and discussed its importance. We introduced the topics of data editing, imputation, and record linkage. We also presented basic techniques for assessing and improving data quality. We are now ready in Part Two of the book to present several specialized data quality techniques. These may, at some level, be familiar to many.

This chapter reviews a few mathematical preliminaries that are necessary for easy understanding of the material in later chapters. Our focus is on three statistical concepts. The first is the concept of conditional independence which we use in the record linkage material of Chapter 8. The second is the Bayesian paradigm of statistics, which has direct application in the Belin–Rubin model, also discussed in Chapter 8. The third topic is the capture–recapture method. This can be used to estimate (1) the number of duplicate entries on a list, (2) the number of duplicate records within a database, or (3) the number of individuals missed in a population census. As a by-product of the discussion on capture–recapture methods, we introduce the concept of a multi-dimensional contingency table. This has application in the edit/imputation material discussed in Chapter 7.

6.1. Conditional Independence[1]

Definition: Two events A and B are said to be *independent* if $P(A \cap B) = P(A)P(B)$.

Definition: Let $\{A_i, i \in I\}$ where I is an arbitrary index set, possibly infinite, be an arbitrary collection of events. The collection of events, $\{A_i, i \in I\}$, is said to be *independent* if for each finite set of distinct indices $i_1, \ldots, i_k \in I$, we have

$$P\left(A_{i_1} \cap A_{i_2} \cap \cdots \cap A_{i_k}\right) = P\left(A_{i_1}\right) P\left(A_{i_2}\right) \cdots P\left(A_{i_k}\right).$$

---

[1] This section is based heavily on pages 26–27 of Ash [1970].

Example 6.1: Let two fair dice be tossed. Let each possible outcome have (an equal) probability of occurrence of $\frac{1}{36}$. Let

$$A = \{first\ die = 1, 2,\ or\ 3\}$$
$$B = \{first\ die = 3, 4,\ or\ 5\}$$
$$C = \{\text{the sum of the two faces is } 9\} = \{(3, 6), (4, 5), (5, 4), (6, 3)\}$$

Hence,

$$A \cap B = \{(3, 1), (3, 2), (3, 3), (3, 4), (3, 5), (3, 6)\},$$
$$A \cap C = \{(3, 6)\},$$
$$B \cap C = \{(3, 6), (4, 5), (5, 4)\},\ \text{and}$$
$$A \cap B \cap C = \{(3, 6)\}.$$

So, it follows that

$$P(A \cap B) = \frac{1}{6} \neq P(A)P(B) = \left(\frac{1}{2}\right)\left(\frac{1}{2}\right) = \frac{1}{4}$$

$$P(A \cap C) = \frac{1}{36} \neq P(A)P(C) = \left(\frac{1}{2}\right)\left(\frac{4}{36}\right) = \frac{1}{18}$$

$$P(B \cap C) = \frac{1}{12} \neq P(B)P(C) = \left(\frac{1}{2}\right)\left(\frac{4}{36}\right) = \frac{1}{18}.$$

Consequently, even though

$$P(A \cap B \cap C) = \frac{1}{36} = P(A)P(B)P(C) = \left(\frac{1}{2}\right)\left(\frac{1}{2}\right)\left(\frac{1}{9}\right),$$

the events $\{A, B, C\}$ are not independent.

Definition: Two events A and B are said to be *conditionally independent given event M* if $P(A \cap B | M) = P(A|M)P(B|M)$.

Definition: Let $\{A_i, i \in I\}$ where I is an arbitrary index set, possibly infinite, be an arbitrary collection of events. The collection of events, $\{A_i, i \in I\}$, is said to be *conditionally independent given event M* if for each finite set of distinct indices $i_1, \ldots, i_k \in I$, we have

$$P\left(A_{i_1} \cap A_{i_2} \cap \cdots \cap A_{i_k} | M\right) = P\left(A_{i_1} | M\right) P\left(A_{i_2} | M\right) \cdots P\left(A_{i_k} | M\right).$$

6.2.   Statistical Paradigms

The two major statistical paradigms for constructing statistical models are: (i) the *frequentist* or *classical* paradigm and (ii) the *Bayesian* paradigm.

In the frequentist paradigm, the probability of an event is based on its relative frequency. All prior and/or collateral information is ignored. Proponents of the frequentist paradigm view it as being objective, because all attention is devoted to the observations (data). Some of the key constructs of the frequentist paradigm are the Neyman–Pearson Lemma, tests of statistical significance, confidence intervals, and unbiased estimates.

In the Bayesian paradigm, probability is treated as a rational measure of belief. Thus, the Bayesian paradigm is based on personal or subjective probabilities and involves the use of Bayes' theorem.[2] Prior and/or collateral information is incorporated explicitly into the model via the prior distribution. Some of the key constructs of the Bayesian paradigm, in addition to Bayes' theorem itself, are conditional probabilities, prior distributions, predictive distributions, and (posterior) odds ratios.

As illustrated by the following simple example, the Bayesian statistician computes his/her probabilities given (i.e., conditional upon) the observed data.

Example 6.2: Bayesian Probabilities

One of two coins is selected with equal probability of .5. The first coin has one side "heads" and the other side "tails." The second coin has both sides "heads." The selected coin is tossed six times. All six outcomes are "heads." What is the probability that the coin selected is the one having both sides "heads"?

Solution:

Let I denote the event that the coin selected has one "head" and let II denote the event that the coin selected has two "heads." We seek P[II | six heads observed]. By Bayes' theorem and the theorem of total probability, we have

$$P[\text{II} \mid \text{six heads observed}] = \frac{P[\text{six heads observed} \mid \text{II}] \cdot P[\text{II}]}{P[\text{six heads observed}]}$$

$$= \frac{P[six\ heads\ observed \mid II] \cdot P[II]}{P[six\ heads\ observed \mid II] \cdot P[II] + P[six\ heads\ observed \mid I] \cdot P[I]}$$

$$= \frac{1 \cdot (.5)}{1 \cdot (.5) + (.5)^6 \cdot (.5)} = \frac{64}{65}.$$

So, because we have observed six heads, the probability that we have selected the coin with heads on both sides has increased from our initial estimate of $\frac{1}{2}$ to $\frac{64}{65}$.

---

[2] Bayes' Theorem states that if A and B are events such that $P[B] > 0$, then $P[A|B] = \frac{P[B|A] \cdot P[A]}{P[B]}$.

6.3.   Capture–Recapture Procedures and Applications

Methods known as capture–recapture procedures have applications to a number of the problems that we have considered earlier. The simplest version of this methodology, also known as *dual systems estimation*, involves two separately compiled, but incomplete, lists of the members of a population. With three or more such lists, the methodology is called *multiple systems estimation*.

Applications of capture–recapture procedures go back to at least 1896, to the work of Petersen [1896] who was interested in the size of fish populations. Another early paper by Lincoln [1930] was concerned with waterfowl. The method gets its name from such applications concerned with estimating the number of fish in a pond, or more generally, estimating the population size of various species in the wild.

For example, we could take a sample of fish in a pond and tag each of the fish so selected. A day later, we could take a second sample of fish from the pond, and count the number of fish in this second sample that had been tagged the day before. Then, using the methods of this section, we could estimate the total number of fish in the pond.

Another classic paper, by Sekar and Deming[3] [2004], was concerned with estimating birth and death rates in an area near Calcutta, India. Capture–recapture procedures also have application to estimating (1) the size of the undercount in censuses and (2) the number of duplicate records on a list/database. These are two of the specific applications we consider in depth in this work. Other possible applications include (1) estimating the number of drug addicts in the United States and (2) estimating the number of children in Massachusetts possessing a specific congenital abnormality. We discuss the general methodology first before considering specific applications. Despite their age, the two most comprehensive treatments of the method may still be the books by Marks, Seltzer, and Krotki [1974] and Bishop, Fienberg, and Holland [1975].

6.3.1.   General Approach to the Two-Sample Capture–Recapture Problem

Let N be the total number of individuals in the population of interest. Let x_{11} denote the number of individuals observed to be in both samples. Let x_{12} denote the number of individuals observed to be in the first sample but not the second; and let x_{21} denote the number of individuals observed to be in the second sample but not the first. Finally, we let $x_{1+}(=x_{11}+x_{12})$ and $x_{+1}(=x_{11}+x_{21})$ denote the number of individuals present in the first and second samples, respectively. We summarize the observations (Table 6.1) in a two-by-two contingency table with one missing value denoted by x_{22}:

---

[3] This paper is an excerpt of Sekar and Deming [1949] and is introduced by Scheuren [2004] as part of his "History Corner" in *The American Statistician*.

TABLE 6.1. Basic 2-by-2 contingency table

| | Second sample | | |
First sample	Present	Absent	Total
Present	x_{11}	x_{12}	x_{1+}
Absent	x_{21}	x_{22}	
Total	x_{+1}		N

The goal here is to estimate the missing value, x_{22}, which leads easily to an estimate of the total population, N. The traditional estimator of N is

$$\hat{N} = x_{11} + x_{21} + x_{12} + \frac{x_{12}x_{21}}{x_{11}}.$$

This estimator can be shown to be equal to

$$\hat{N} = \frac{x_{1+}x_{+1}}{x_{11}}.$$

and can be derived by assuming the following identity holds:

$$x_{11}x_{22} = x_{12}x_{21}.$$

(This assumption holds when the two samples are independent.) The estimator, $\hat{N}$, is sometimes called the *Lincoln–Peterson estimator* in which case we could rewrite it as $\hat{N}_{LP}$.

Such dual systems estimators rely on three assumptions. The first assumption is that the samples or lists are independent. In other words,

$$P[item\ i\ on\ list\ L_1 \mid item\ i\ on\ list\ L_2] = P[item\ i\ on\ list\ L_1].$$

The second assumption is that the population of interest is homogeneous in the sense that each member of the population has an equal chance of being captured *for a given list*.

The third assumption is that there are no errors when matching items across lists. Moreover, an accurate estimate of the number of items in both lists, x_{11}, is particularly crucial to this process.

Before considering some examples of the above, we want to make two additional comments. First, most, if not all of this work, should be automated because clerical matching is too slow and too prone to produce errors. We next summarize some comments of Sekar and Deming [2004] regarding the assumption of independence.

Sekar and Deming [2004, pp. 14–15] argue that it is important to realize that "correlation signifies heterogeneity in the population [because] it implies that events that fail to be detected do not form a random sample of the whole population of events. This heterogeneity may arise only if there are differences in reporting rates among different segments of the population, resulting in the group of failures being weighted disproportionately by the different segments."

"It therefore follows that the correlation can be minimized" by (1) partitioning "the population into homogeneous groups" and then (2) "calculating the total number of events separately for each group." The grand total can then be obtained by addition. For more details, the interested reader should see Sekar and Deming [2004].

6.3.2. Weevils in a Box of Wheat

As Bishop, Fienberg, and Holland [1975] report, Andrewartha [1961] describes an experiment in which about 2,000 weevils were placed in a box of wheat and allowed to disperse. An initial sample of 498 weevils was drawn without replacement from the box of wheat. Each of the weevils selected was marked with a small spot of paint and put back into the box of wheat. A week later, a second sample of 110 weevils was drawn without replacement. Twenty of those selected in the second sample were observed to have a spot of paint. Using the capture–recapture scheme described above, we obtain an estimate of

$$\hat{N}_{LP} = \frac{x_{1+}x_{+1}}{x_{11}} = \frac{498 \cdot 110}{20} = 2,739$$

for the total number of weevils in the box.

6.3.3. Estimating Birth and Death Rates in India

Sekar and Deming [2004] describe the results of a study conducted during February 1947, in an area known as the Singur Health Centre, near Calcuttta, India. The total area of the Centre is about 33 square miles. At the time of this study, the total population of the area was about 64,000 people living in about 8,300 houses.

Sekar and Deming use capture–recapture methods to estimate the number of births and deaths for residents of this area during each of the years 1945 and 1946.

The headman of each village periodically submits a list of births and deaths to a registrar. The registrar coordinates this information with a second report from each village and a list from the Maternity and Child Welfare Department. We refer to the resulting list as the "registrar's list of births and deaths" – the "R-list."

During an 11-week period beginning on February 11, 1947, interviewers from the All-India Institute of Hygiene and Public Health visited every house within the Singur Health Centre to prepare a list of all of the births and deaths that occurred during 1945 and 1946 – the I-list.

After deleting the non-verifiable, illegible, incomplete, and incorrect items from the R-list, Sekar and Deming applied the two-sample technique described above. We summarize these corrected data in Table 6.2, a table similar to Table 6.2-1 in Bishop, Fienberg, and Holland. Table 6.2-1 is an abbreviated version of Table I on page 108 of the original Sekar and Deming article.

TABLE 6.2. The investigators' report on the comparison of the lists of the Singur health centre

| Event[4] | Year | Total | Registrars' lists | | Interviewers' lists |
			Found in interviewer's Lists	Not found in interviewer's lists	Extra in interviewers' lists
Births	1945	1,504	794	710	741
	1946	2,242	1,506	736	1,009
Deaths	1945	1,083	350	733	372
	1946	866	439	427	421

To illustrate the capture–recapture methodology, we consider the deaths occurring during 1946. We summarize (Table 6.3) the data of interest in a 2-by-2 table:

TABLE 6.3. Number of deaths occurring during 1946 in the Singur health centre

| R-List | I-List | |
	Present	Absent
Present	439	427
Absent	421	?

The estimate of the total number of deaths occurring during 1946 is

$$\hat{N}_{LP} = \frac{(439+421) \cdot (439+427)}{439} = 1,696$$

where we have rounded the result to the nearest integer.

Finally, for computing birth and death rates in this area, the population base assumed was that furnished by the household interviews.

6.3.4. Estimating the Number of Duplicates within a Database

Sometimes, we do not have a list of all the records within a database that have a particular type of error – for example, an incorrect Zip Code. By taking several (independent) samples of records from this database and counting the number of records in each overlapping sample with an erroneous Zip Code, we can obtain an efficient estimate of the number of records in the entire database with

---

[4] The events referred to here are those listed as occurring in the village that did not involve institutionalized non-residents.

an erroneous Zip Code. This procedure is frequently very efficient because we only need to sample a small proportion of the records in the entire database to produce our estimate.

Here, we use capture–recapture to estimate the number of duplicate records within a database. In Chapter 14, we use capture–recapture to estimate the number of records having a certain type of error associated with the editing process. In all situations, we use small subsets of the population of interest that contain much higher proportions of erroneous records than would a simple random sample of the entire population. This is highly efficient in terms of use of resources.

Example 6.3: Database on Credit Card Applicants

A database consisting of information on individuals applying for a credit card contained the following six data elements on these individuals: Last Name, First Name, Middle Name, Social Security Number, Date of Birth, and Place of Birth. Of these, 498 pairs of records having identical Social Security Numbers were determined to be duplicates (after extensive review) while 110 pairs that agreed exactly on Last Name, First Name, Middle Name, Date of Birth, and Place of Birth were similarly determined to be duplicates. Finally, 20 pairs of records that agreed on all six data elements were determined to be duplicate records. (This is the number of duplicates in common between the two matching schemes.) How many duplicate records were to be found in the entire database?

The answer is 2,739. This is the answer to the problem posed in Section 6.3.2, as the equations in both examples are identical. We note that the approach taken in this example is very conservative because the matching criteria used are extreme in the sense that records having minor typographical errors in their name fields, for example, may not be classified as duplicate records. In Chapter 8, we describe a less extreme approach to this type of problem.

6.3.5. *Killings and Refugee Flow in Kosovo*

Ball, Betts, Scheuren, Dudokovic, and Asher [2002] estimated the number of people killed in Kosovo during the spring of 1999 (Table 6.4). They had four separate source lists of individual victims of killing:

- American Bar Association Central and Eastern Europe Law Initiative (ABA/CEELI),
- Exhumations (EXH),
- Human Rights Watch (HRW), and
- Organization for Security and Cooperation in Europe (OSCE).

The data of interest are summarized in the following $2 \times 2 \times 2 \times 2$ contingency table.

While there is essentially only one method of estimation – the basic (Lincoln–Peterson) estimator – in the case of two lists, there is much more flexibility in the case of three or more lists. Both Bishop, Fienberg, and Holland [1975] and

TABLE 6.4. Number of individual victims of killing by documentation status (including victims with imputed dates of death)

HRW	ABA EXH OSCE	Yes Yes	Yes No	No Yes	No No	Total
Yes	Yes	27	32	42	123	
Yes	No	18	31	106	306	
No	Yes	181	217	228	936	
No	No	177	845	1,131	??	
Total						4,400

TABLE 6.5. Dual system estimates

	EXH	HRW	OSCE
ABA	7,242	9,691	5,972
EXH		6,779	7,137
HRW			5,462

TABLE 6.6. Three-way and four-way system estimates (under saturated model)

Lists used	Estimated population total
ABA, EXH, HRW	11,811
ABA, EXH, OSCE	22,301
ABA, HRW, OSCE	12,255
EXH, HRW, OSCE	8,016
ABA, EXH, HRW, OSCE	16,942

Marks, Seltzer, and Krotki [1974] describe a variety of estimators that can be used in such a multiple systems situation. The estimates we computed are given in Tables 6.5 and 6.6. The estimates of the number of victims based on the use of three of the four lists are computed using equations 6.4–14 and 6.4–15 on p. 239 of Bishop, Fienberg, and Holland [1975]. The estimate based on the use of four lists is computed using Equation 14 on page 41 of Ball, Betts, Scheuren, Dudokovic, and Asher [2002]. We note that our estimates are in most cases only slightly different from those reported in Ball et al. The exception is that our estimate based on all four lists is considerably higher than theirs.

6.3.6. Further Thoughts on Capture–Recapture

- An assumption underlying the Lincoln–Peterson estimator is that the probabilities of being on the list frames are homogeneous (i.e., identical). Haines, Pollock, and Pantula [2000] extend the results to the case where the inclusion probabilities are heterogeneous (i.e., unequal). For example, larger farms may have a greater probability of being on a list frame than smaller farms. They also show how the inclusion probabilities can be modeled as functions of auxiliary variables using a logistic regression model. For example, in capture–recapture work, the probabilities of inclusion (i.e., capture) may vary with the age, gender, or size of the species being studied.

- When neither the homogeneity assumption nor the independence assumption is valid, then capture–recapture can produce substantially biased estimates in two-list capture–recapture settings. When the homogeneity assumption is valid and three or more lists are used, then the information from the extra lists may compensate for the lack of independence (Zaslavsky and Wolfgang 1993; Winkler 2004) in the first two lists and yield relatively unbiased capture–recapture estimates. Because of the (exceptionally) small proportion of true matches within the large set of pairs from two or more lists, simple random sampling will not yield unbiased estimates of the proportion (or number) of true matches that are not found by a set of blocking criteria (Deming and Gleser [1959], Winkler [2004, ASA-SRMS]).

- Finally, in the Lincoln–Peterson approach, it is possible that the denominator, x_{11}, is zero. Chapman [1951] has proposed an estimator

$$\hat{N}_{CH} = \frac{(x_{1+}+1)(x_{+1}+1)}{(x_{11}+1)} - 1$$

that circumvents this difficulty. The Chapman estimator also leads to a reduction of the bias in the estimation process. Of course, if the denominator of the estimator is close to zero, the estimates will be unstable in the sense that small changes in the value of the denominator will lead to large changes in the value of the desired estimate. For more details on the bias of such estimators, the reader should see Sekar and Deming [2004] and Chapman [1951].

7
Automatic Editing and Imputation of Sample Survey Data

7.1. Introduction

As discussed in Chapter 3, missing and contradictory data are endemic in computer databases. In Chapter 5, we described a number of basic data editing techniques that can be used to improve the quality of statistical data systems. By an *edit* we mean a set of values for a specified combination of data elements within a database that are jointly unacceptable (or, equivalently, jointly acceptable). Certainly, we can use edits of the types described in Chapter 5.

In this chapter, we discuss automated procedures for editing (i.e., cleaning up) and imputing (i.e., filling in) missing data in databases constructed from data obtained from respondents in sample surveys or censuses. To accomplish this task, we need efficient ways of developing statistical data edit/imputation systems that minimize development time, eliminate most errors in code development, and greatly reduce the need for human intervention. In particular, we would like to drastically reduce, or eliminate entirely, the need for humans to change/correct data. The goal is to improve survey data so that they can be used for their intended analytic purposes. One such important purpose is the publication of estimates of totals and subtotals that are free of self-contradictory information.

We begin by discussing editing procedures, focusing on the model proposed by Fellegi and Holt [1976]. Their model was the first to provide fast, reproducible, table-driven methods that could be applied to general data. It was the first to assure that a record could be corrected in one pass through the data. Prior to Fellegi and Holt, records were iteratively and slowly changed with no guarantee that any final set of changes would yield a record that satisfied all edits.

We then describe a number of schemes for imputing missing data elements, emphasizing the work of Rubin [1987] and Little and Rubin [1987, 2002]. Two important advantages of the Little–Rubin approach are that (1) probability distributions are preserved by the use of defensible statistical models and (2) estimated variances include a component due to the imputation.[1]

---

[1] We do not discuss procedures for calculating estimated variances in the presence of missing data. Instead, we refer the reader to Little and Rubin [2002]. However, such estimates are crucial in deciding when a sufficient amount of editing has been done.

In some situations, the Little–Rubin methods may need extra information about the non-response mechanism. For instance, if certain high-income individuals have a stronger tendency to not report or misreport income, then a specific model for the income-reporting of these individuals may be needed.[2] In other situations, the missing-data imputation can be done via methods that are straightforward extensions of hot-deck. We provide details of hot-deck and its extensions later in this chapter.

Ideally, we would like to have an all-purpose, unified edit/imputation model that incorporates the features of the Fellegi–Holt edit model and the Little–Rubin multiple imputation model. Unfortunately, we are not aware of such a model. However, Winkler [2003] provides a unified approach to edit and imputation when all of the data elements of interest can be considered to be discrete. The method can be extended to continuous data by partitioning the continuous data into (sometimes large) sets of intervals. This effectively transforms continuous variables into discrete ones. Although there are no standard, general methods for modeling the joint distributions of continuous data, standard loglinear modeling methods (see, for example, Bishop, Fienberg, and Holland [1975]) can be used with discrete data. If the intervals are made sufficiently small, then *discretized* versions of the continuous variables can often approximate the original continuous variables to any desired degree of accuracy.[3]

The main issue in the unified model is providing good joint distributions that satisfy the logical constraints of the edit rules. As an example, we would not impute a marital status of married to a child of fifteen in the situation where an edit constraint did not allow that particular combination of age and martial status. Within a set of edit constraints, we also want to preserve the joint distributions in a manner that is consistent with the ideas of Little and Rubin.

After presenting ideas on edit and imputation, we then discuss some of the software that has been developed to perform automatic editing and conclude with some thoughts on the appropriate amount of time and money to devote to editing sample surveys and censuses.

Many examples are from the survey-sampling literature because many of the ideas were originally introduced in statistical agencies. Because of the efficiency and power of the methods, they are just beginning to be investigated in other environments such as general business accounting systems and administrative systems. In such applications, these methods may be useful in preventing and identifying data problems. However, some analysts may want to attempt to correct all of the errors in some business databases – for example, corporate billing systems – because of the financial uses of such databases.

----

[2] For an example, see Kennickell [1999].
[3] When sampling zeroes arise, there are limitations on the degree of accuracy that can be attained. We avoid recommending a single approach to modeling sampling zeroes, although one approach is simply to treat them as population zeroes.

The generalized methods based on the ideas of Fellegi and Holt and of Little and Rubin are easily adapted to other situations. The key concept is the preservation of joint distributions in a principled manner so that the databases can be data mined and used in statistical analyses to produce valid inferences.

7.2.   Early Editing Efforts

Data editing has been done extensively in the national statistical agencies. Early work was primarily done clerically. Sometimes experienced interviewers made changes during an interview. In other situations, clerks would be given a set of rules that told them how to "correct" a set of survey forms. After the clerical corrections, the data would be keyed. To increase efficiency and still make use of the subject-matter expertise of the edit rules, later editing methods applied computer programs that mimicked the "if-then-else" rules that clerks had used in their manual reviews.

These automated operations had three advantages:

- Such software could edit more forms than humans could.
- The software performed these edits in a consistent fashion, whereas there were usually inconsistencies in the edits among the clerks performing these edits manually.
- The software eliminated many errors that the clerks failed to correct.

Unfortunately, these methods had three principal disadvantages.

- First, it was difficult to write software that could effectively implement the if-then-else rules. There were often hundreds, if not thousands, of such rules. Sometimes the written specifications for the edits required hundreds of pages. It was extremely difficult to create suitable test decks (i.e., test data) that would effectively test the implementation of the bulk of these if-then-else rules. The debugging of the code was difficult and still did not ensure that the code would perform according to specification during production editing runs.
- Second, there was still a lot of clerical review. Although the majority of records was edited only by the computer software, a large number of the records was still edited manually as an additional quality step. For example, in a large economic census, the clerical review and callbacks might involve 10,000 records – 0.5% of 2 million records. For some surveys with complex edit patterns, the clerical review entailed months of effort by dozens of clerks, thereby delaying the release of both the tabulated data and the public-use files.
- Third, the edit procedures could not ensure that a record would pass edits after a single review by the clerks. Typically, a number of iterations were needed until a record that failed the initial edits passed all of the required edits. As fields were changed to correct for edit failures, additional edits would fail that

had not failed originally. The sequential, local nature of the edit corrections needed multiple iterations to ensure that the "corrected" record would pass all required edits. During this iterative process, more and more fields would be changed on the edit-failing records. In a dramatic demonstration, Garcia and Thompson [2000] showed that a group of 10 analysts took 6 months to review and correct a moderate size survey that had complex edit patterns. In contrast, an automated method based on the model of Fellegi and Holt (see the next section) needed only 24 hours to edit/impute the survey data and only changed 1/3 as many fields as the clerical review.

7.3. Fellegi–Holt Model for Editing

Fellegi and Holt [1976] provided a theoretical model for editing. This model has the virtue that, in one pass through the data, (1) an edit-failing record can be modified so that it satisfies all edits and (2) the logical consistency of the entire set of edits can be checked prior to receipt of data. All edits were in easily maintained tables rather than in thousands of lines of if-then-else rules. The main algorithm for determining what fields to change was implemented in a generalized computer program that could be easily applied in a wide variety of circumstances.

Fellegi and Holt had three goals, the first of which is known as *the error localization problem*:

1. The data in each record should be made to satisfy all edits by changing the fewest possible items of data (variables or fields).
2. Imputation rules should be derived automatically from edit rules.
3. When imputation is necessary, it is desirable to maintain the marginal and joint distributions of variables.[4]

---

[4] Fellegi and Holt, in fact, demonstrated that it is possible to devise a method for filling in missing values or replacing contradictory (i.e., inconsistent or invalid) values in a manner that ensures the "corrected" record satisfies all edits. Due to the global optimization of their method, they showed that it is always possible to delineate a set of fields that, when changed, would yield a record that satisfied all edits. This was very different from prior editing schemes based on if-then-else rules. Under these prior schemes records would still fail some of the edits after passing through the main logic of the software. While the Fellegi and Holt work established the existence of such a solution, their 1976 paper did not deal with many of the computational issues required to obtain an explicit solution. Although the original Fellegi-Holt model appeared in the statistical literature, the algorithms used to implement their model have primarily relied on optimization methods from the operations research (OR) literature. Specifically, integer-programming, an OR optimization procedure, is used to find the minimum number of fields to impute on each record so that the resulting record satisfies all edits. The specifics of integer-programming approaches are beyond the scope of this work.

7.4.   Practical Tips

In order to construct a suitable set of edits or business rules, it is often useful, if not indispensable, to obtain assistance from subject matter experts. They have the best knowledge of how the data are collected, what errors that they have observed, and how the data are used.

Identify extreme data values. The presence of such outliers can be an indicator of possible errors. Make use of (1) simple univariate detection methods as Hidiroglou and Berthelot [1986] suggest and/or (2) the more complex and graphical methods that de Waal [2000] suggests. Two interesting applications in a survey sampling context are Bienias, Lassman, Scheleur, and Hogan [1995] and Lu, Weir, and Emory [1999].

Follow up selectively. Latouche and Berthelot [1992] suggest using a score function to concentrate resources on (1) the important sample units, (2) the key variables, and (3) the most severe errors. Moreover, this allocation of resources may depend upon how the data will be used.

As discussed in Section 2.3.4, place more emphasis/resources on error prevention than on error detection. To this end, move the editing step to the early stages of the survey process, preferably while the respondent is still available. If necessary, try to use computer-assisted telephone or personal or self-interview methods to recontact the respondent.

Editing is well-suited for identifying fatal errors because the editing process can be so easily automated. Perform this activity as quickly and as expeditiously as possible. Wherever possible employ generalized, reusable software.

Certain automatic corrections to satisfy edit or business rules may be built into software that is run prior to loading records onto databases. (These are sometimes called *front-end* edits.) Ideally, all, or almost all, "editing" can be done automatically so that little, or no, follow-up is required. If too high a proportion of data fails edits and does not get placed in a database, then it may not be feasible to use the data for the purposes intended. For example, a database with lots of errors may preclude a chain of department stores from using its data for customer relations and marketing.

As the following example shows, different edit rules may be needed according to how the data will be used.

Example 7.1: Limited editing for tabulated data

A statistical agency may only need to edit data sufficiently well to assure that it can be used to produce tabulations in publications. Some tabulations will be totals over the entire database; others will be restricted to subdomains such as those determined by State codes or by certain age ranges. If the totals are by State codes, then the State codes may need to be edited and corrected. If the totals are by certain age ranges, then it is possible that an unusual tabulation in an age range (say age greater than 80) may need to have the ages more carefully edited. For a total over the entire database, an erroneous State code or a set of erroneous ages may be of little or no consequence.

The number of edit rules can grow according to the specific uses of a database. In the above examples of simple totals, we might need modest sets of edits. As the number of analyses grows, then we might need a large number of edits. In an ideal world, perfect data or "edited" data would represent some underlying reality. The data could be used for all analyses. If we know the specific analyses for which the data will be used, then we can delineate certain aggregates (such as first and second moments or sets of pairwise probabilities) that need to be accurate. In those situations, we can target only the most important records involved with the aggregates. If there are a moderate number of aggregates, we can reduce the number of edits to only those needed for preserving the aggregates. A record, however, that has an error in one field will often have one or more other errors in other fields.

For example, will the database need to contain several telephone numbers where one might be a typical contact point, one might be a work phone number, and another might be the phone number that is associated with a current credit card?

7.5.   Imputation

In this section, we describe and then compare a number of approaches to imputation that have been previously applied. Such approaches include hot-deck, cold-deck, and model-based approaches.

7.5.1.   Hot Deck

Little and Rubin [1987] broadly define *hot-deck imputation* "as a method where an imputed value is selected from an [assumed or] estimated distribution for each missing value, in contrast to *mean imputation* where the mean of the distribution is substituted. In most applications, the distribution consists of values from responding units, so that the hot deck imputation involves substituting individual values drawn from similar responding units." The name "hot deck" has its roots in the early days of computers when the non-respondent's imputed value was selected from a deck of computer cards of matching donors. In the original US Census Bureau application, the matching donor resided in the same geographic region and had the same household size as the non-respondent. For more information on hot-deck procedures, the reader should see Ford [1983], Oh and Scheuren [1980, 1983], or Oh, Scheuren, and Nisselson [1980].

In the most elementary situation in which we assume that the non-respondents are distributed randomly among a population, we can choose a respondent at random and substitute that respondent's value for the missing value of the non-respondent. The repetition of this procedure for each missing value is a mechanism for filling in all of the missing values. A more sophisticated procedure involves matching the respondent and non-respondent on one or more characteristics (or background variables). We illustrate this in the following example.

Example 7.2: Predicting appraised values in a survey of houses

A household survey is conducted annually to obtain information on the characteristics of houses within a given county. The items collected include the number of bedrooms, the number of bathrooms, and the most recent appraised value of the house. For the records on a few houses included in the survey conducted during calendar year 2005, the appraised value was missing. A hot-deck imputation scheme was used to impute the missing appraised values based on the number of rooms and the number of bedrooms. For one house, "A," whose record had a missing appraised value, the number of reported bedrooms was four and the number of reported bathrooms was two. The database for this survey had 15 records that contained an appraised value and also reported four bedrooms and two bathrooms. One of these fifteen records was chosen at random and the reported appraised value of $195,000 on that house was then used as the imputed value for house "A."

Example 7.3: Too many bathrooms?

For another house, "B," which also had a missing appraised value, the number of reported bedrooms was eight and the number of bathrooms reported was five. There were no other houses in the survey with this many bedrooms or bathrooms. This presented a problem for the survey manager. A decision was made to impute the appraised value of the record with the next largest number of bedrooms – namely six – even though it was felt that this might induce some bias into the results of the survey.

If two or more fields on a record have missing values, then all of the missing values can be substituted from an appropriate single matching donor record. This hot-deck approach typically does a better job of preserving joint distributions than a scheme in which each missing value is taken from a separate donor record.

Most hot-deck estimators are unbiased only under the unrealistic assumption that the probability of response is not related to (the value of the) response itself. If additional (i.e., covariate) information is available on both the responding units and the non-responding units, then this information may be used to reduce the bias resulting from the non-response. Little and Rubin [1987] describe two such approaches.

They call the first scheme a *hot deck within adjustment cells*. Under this scheme, the analyst partitions both the respondent and non-respondents according to the background variables or covariates that the analyst deems appropriate. (Little and Rubin [2002, Example 4.8] offers some suggestions on how to do this.) Each component of the partition is known as a *cell*. The mean and variance of the resulting hot-deck estimates of the statistics of interest can be computed by first using the customary hot-deck method separately for each cell and then combining the results over all of the cells of the partition. Little and Rubin [2002] point out that this procedure may not work well for continuous quantitative variables because such variables would have to be partitioned into a small number of intervals.

Little and Rubin [2002, Example 4.9] call the second scheme a *nearest neighbor hot deck*. Under this approach, the analyst first defines a metric that measures the distance between sampling units, based on the values of the units' covariates. The second step is to select imputed values from the responding units that are near the unit with the missing value. For example, let $x_{i1}, x_{i2}, \ldots, x_{iJ}$ be the values of J appropriately scaled covariates for a unit i for which y_i is missing. Little and Rubin define the distance between units i and i' as

$$d(i, i') = \max_{1 \leq j \leq J} |x_{ij} - x_{i'j}|.$$

Then, under a nearest neighbor hot-deck, an imputed value for y_i is selected from one of the units i' for which (1) $y_{i'}, x_{i'1}, \ldots, x_{i'J}$ are all observed and (2) $d(i, i')$ is less than some value d_0. The number of possible donor units, i', is determined by the choice of d_0.

Typically, the jth covariate, x_{ij}, is scaled by subtracting its sample mean and dividing by its sample standard deviation. This causes each covariate to have the same relative weight in $d(i, i')$.

If the covariates are not scaled and one of the covariates has a wide range of values, then that covariate may dominate the other variables in the distance function $d(i, i')$.

An indicator function is typically used as the distance function for a covariate that only assumes non-numeric values. Here exact agreement is assigned a value of 0 and disagreement is assigned a value of 1.

Example 7.4

In the survey described in Example 7.2, the ith unit might be the one with the missing appraised value, y_i. We could use x_{k1} to denote the number of bedrooms of the kth unit in the survey and x_{k2} to denote the number of bathrooms of the kth unit in the survey. If, for a nearest neighbor hot deck, we chose $d_0 = 0$, then we would only select donor units from those units that agreed exactly on both the number of bedrooms (i.e., 4) and the number of bathrooms (i.e., 2). Alternatively, we might choose $d_0 = 1$. In this case, we would pick a donor unit from those units that had 3, 4, or 5 bedrooms and between 1 and 3 bathrooms.

7.5.2.   Cold Deck

In *cold deck imputation* a constant value from an external source, such as a previous realization of the same survey, is used as a substitute for the missing value of a data element.

Example 7.5: Using a cold deck to impute an appraised value

In the household survey of Example 7.2, in order to impute a missing appraised value via a cold-deck imputation scheme, a record on a house with four bedrooms and two bathrooms is chosen from the database constructed for the earlier version of the survey conducted during calendar year 2004.

Values from a cold deck are often used as the imputed values at the start of a hot-deck imputation. The following example, dealing with the Current Population Survey (CPS) and extracted from Ford [1983], illustrates this.

Example 7.6: Using Cold-deck Values to Jump Start a Hot deck

The CPS hot-deck is a sequential procedure in which the ordering is roughly geographic. As the computer processes the sample, the hot-deck values are continually updated to reflect the most recently processed sample cases. When a labor force item is missing, the hot-deck value – a value obtained from the most recently processed record which does not have any missing items – is imputed for the missing item. Because the first record in a sex–race–age cell may have missing values, cold-deck values are stored in the computer before processing the sample. Cold-deck values are based upon the data of the previous CPS surveys.

7.5.3. Model-Based Approach

Here, missing values are replaced by predicted values produced by a statistical model – for example, a multiple linear regression model or a logistic regression model. Sometimes, the value selected may be the mean of the regression model plus an empirical residual value assigned through a hot deck and drawn to reflect the uncertainty in the predicted value. Ideally, we would like to have a statistical model that incorporates all of the data on the complete records as well as the available data on the records that have some missing data elements. We would also like our model to preserve the original joint distributions of the data (as the hot deck does) in a statistically principled fashion – that is, according to a plausible model. (The basis for the hot deck is more heuristic.)

Example 7.7: Using a regression model to impute an appraised value

For the survey described in Example 7.2, we could use the following naïve multiple linear regression model in which (1) the number of bedrooms and (2) the number of bathrooms are used as predictor variables to impute an estimated appraised value

$$appraised\ value = \$55,000 + \$30,000 \cdot (\#of\ bedrooms)$$
$$+ \$10,000 \cdot (\#of\ bathrooms) + \varepsilon.$$

where ε is the random error term.[5]

---

[5] The equation of Example 7.7 is used to illustrate our approach. In practice, the coefficients would typically be estimated from the observed values by the method of least squares.

An advantage of using a regression model is that the imputed values may be closer to the "true underlying values" than those obtained via hot deck. This is particularly true when the set of complete records is not sufficiently large to assign values from a unique donor record to each record having missing values.

7.5.4. Multiple Imputation Models

In *multiple imputation models*, two or more values are substituted for the value of each missing data element. At least two copies of a *completed database* are created. By a *completed database* we mean one in which all missing values are replaced by suitably imputed values.[6]

There is even readily available, general-purpose software for (1) doing multiple imputation and (2) analyzing the results of multiple imputation. For example, SAS® version 8.2 has (1) PROC MI and (2) PROC MIANALYZE.

7.5.5. Comparing the Various Imputation Approaches

An important limitation of single imputation schemes is that standard variance formulas applied to filled-in data systematically underestimate the variance of the estimates, even if the procedure (e.g., the model) used to generate the imputations is correct. The imputation of a single value treats that value as known. Thus, without special adjustments, single imputation cannot reflect the sampling variability correctly. Moreover, the relationship among variables (i.e., the correlation structure) is frequently distorted by the use of a single imputation, especially via a hot deck. Multiple imputation procedures allow valid estimates to be calculated using standard complete data procedures as well as off-the-shelf statistical software. Moreover, multiple imputation procedures facilitate the calculation of the component of the variance due to the imputation process itself.

Although it is straightforward to determine predictive variables using linear regression or logistic regression, such statistical procedures are seldom used. They have been considered time-consuming to use and require high-powered statistical staff with specialized modeling experience/expertise. Instead, hot-deck procedures are used. The predictive variables used in the hot-deck are often chosen based on superficial observation rather than on more rigorous statistical analysis.

Hot-deck schemes are often adversely affected by practical difficulties that the analyst does not anticipate, however. One such potential deficiency is that it implicitly assumes there will be many (possibly thousands) of donors for each given pattern (set of values of predictive variables) on which the matching is to be done. This is almost never the case. See Example 7.3 for just such a situation. Hot-deck schemes also have trouble imputing in constrained environments such as balance sheets, income statements, or income tax returns. Hinkins [1984] reports

---

[6] In practice, we recommend creating at least three completed databases.

that hot-deck imputation procedures applied to corporate income tax returns severely distorted the distributions within micro-datasets for some variables and some subpopulations (e.g., banks). This problem of almost over-constrained imputation settings is common in both (1) administrative or operation record settings and (2) business surveys.

When programmers implement hot decks, they often have collapsing rules for matching on subsets of predictive variables. This collapsing distorts the joint distributions. So, hot-deck procedures usually do not preserve joint probability distributions. In addition, although in many practical situations hot-deck procedures preserve single-variable probability distributions, at least approximately, there is nothing in the hot-deck method that guarantees this will occur. Finally, there is no way to estimate sampling errors when only a single imputation is created.

7.6. Constructing a Unified Edit/Imputation Model

This complex and important subject is a fertile area for research. Ideally, we would like a simple, coherent model that unifies the ideas of both (1) the editing scheme of Fellegi and Holt and (2) the multiple imputation modeling approach of Little and Rubin.

Currently, hot-deck and more general imputation models have not included logical constraints of values that can be imputed. For instance, if the ratio of imputed total wages to observed total hours worked is too large, then we want to impute a smaller value of total wages that satisfies the logical constraint. In other situations, we may not want to impute an age of less than 16 to a person who has a marital status of "married." In the first situation, we might increase the probabilities in certain ranges of total wages or total hours worked to account for the ranges that are unacceptable due to the edit constraints. In the latter situation, we might draw values from a suitable distribution that does not allow ages below 16 for someone who has a marital status of "married." In both situations, the probabilities from the constrained distributions need to be adjusted so that the probabilities add to one.[7]

In simple situations where the data are discrete and the missing data mechanism is missing at random, Winkler [2003] provides a theoretical solution for imputing values that satisfy edit constraints. The main difficulty in practical situations is extending imputation methods to larger situations having more than 8–12 variables. In smaller situations, the methods of Little and Rubin [2002, Section 13.4] are computationally tractable. Their EM-algorithm-based methods[8]

---

[7] Creative, ad hoc schemes can sometimes be used in imputation settings where editing constraints restrict hot-deck imputations. In Mustaq and Scheuren [2005], for example, percentages were imputed to partly overcome this difficulty.

[8] The EM algorithm is a maximum-likelihood based, iterative procedure that Baum, et al. [1970] proposed. Dempster, Rubin, and Laird [1977] developed this scheme independently

for modeling the overall distribution of the data are a special case of more general methods developed by Winkler [1993] for record linkage. The edit methods are currently tractable in nearly all survey situations and some of the larger data warehousing situations. If a specific model is available for the missing data mechanism,[9] then the Little and Rubin method and the Winkler method extend in a straightforward fashion. If continuous data are available, then extensions that partition the continuous data into a set of discrete intervals will work in many situations. The alternative to discrete intervals is to do detailed modeling of the conditional continuous distributions.

The first step in the Winkler [2003] scheme is to summarize the data within a multi-dimensional contingency table. The methods of Fellegi and Holt are then used to edit the data and to determine which cells (i.e., combinations of data elements) of the contingency table are not allowable; such cells are called *structural zeros*. (For example, these might include cells in which an individual's age was "less than 16" and whose marital status was "married.") The next step is to construct a good-fitting, parsimonious log-linear model to represent this contingency table.

The entire set of data (including the records with incomplete data) is fit via EM-algorithm-based methods that are similar to the methods introduced by Little and Rubin [2002, Section 13.4]. A general EM algorithm (Winkler [1993]) is used that accounts for the structural zeros (Winkler [1990]) during the fitting process.

The idea is to fit several different log-linear models with one or more inter-action terms,[10] and to select the best-fitting model. Values are drawn from this model to fill in for the missing data. This is done by matching on the non-missing data elements of the non-respondent's record and then using standard Monte Carlo methods (see, for example, Herzog and Lord [2002]) to draw a value from the log-linear model to replace the missing data element. If only the complete data records are used to build these log-linear models, then general procedures can be found in texts such as Bishop, Fienberg, and Holland [1975].

Little and Rubin [2002, Chapters 13 and 15] describe more sophisticated models that incorporate the data from both the complete and incomplete records. The key difference is that the joint probabilities in the models based on both the complete and incomplete data are different (and more plausible) than the proba-bilities based on models using only complete data. For instance, if individuals in higher-income ranges have higher non-response to income questions and the probability of being in a given response range can be reasonably predicted (i.e.,

---

and popularized it. For more details on the EM algorithm, the interested reader should see Section 6.6 of Hogg, McKean, and Craig [2005].

[9] If the units with unobserved values are a random subsample of the sampled units, then the missing data are *missing completely at random*. If the units with the unobserved values are a random sample of the sample units within each non-response cell, then the missing data are *missing at random*. If the probability that y_i is observed depends on the value of y_i, then the missing-data mechanism is *"non-ignorable* and analyses [that are based] on the reduced sample" and that do not take this into account "are subject to bias."

[10] Bishop, Fienberg, and Holland [1975] have an extended discussion of such models.

imputed) using other reported values, then the more sophisticated models will more appropriately adjust the higher-income ranges upward whereas the models based on only complete data will not.

The Little–Rubin [2002] model from Chapter 13 extends hot-deck imputation in a manner that preserves joint distributions. Hot-deck imputation typically does not preserve joint distributions. The model from Chapter 15 extends the ideas to non-ignorable non-response. If the higher-income individuals have higher non-response rates and the values of the higher-income individuals cannot be effectively predicted from other reported values, then a separate model for the response propensity of the higher-income individuals is needed. In all situations, the results of the modeling are probability distributions that represent the entire population of individuals. Again, we use Monte Carlo methods to draw values from the probability distributions to obtain values for the missing data.

If the imputation procedure is performed at least twice for each missing data element (preferably using different models each time), then the component of the mean square error due to the imputation process can be computed. Moreover, this procedure preserves the joint statistical distributions of the underlying data in a way that is consistent with both the observed marginal totals of the contingency table and the interaction terms used in the log-linear model.[11]

With classical hot-deck, it is typical to have no more than one donor for each missing-data record. If there is one donor, then we are unlikely to effectively estimate the component of the mean square error due to the imputation process. If there are no donors, then the matching is generally only done on a subset of the reported variables in a record. In this situation, joint distributions are likely to be severely distorted.

7.7. Implicit Edits – A Key Construct of Editing Software

A key construct of many current state-of-the-art automated editing systems is a concept known as an "implicit edit." An implicit edit is an edit that is logically implied by two or more explicit edits. Explicit edits are usually specified prior to the development of an editing system. The implicit edits contain information about explicit edits that may fail as a record is changed; such information is used to enable the record to pass all explicit edits. This is the key difference between the model of Fellegi and Holt and earlier models. The earlier models could not guarantee that a record could be corrected in one straightforward pass. Generally, complicated, iterative procedures were used and still could not guarantee that a record would be "corrected" in the sense of no longer failing an explicit edit.

---

[11] While the multiple imputation process has attractive features, the missing data items remain missing; so, in order to aid the analysts of the database, the imputed values must be identified as such in the database.

A number of researchers have shown that various direct integer-programming methods of solving the error localization problem (i.e., finding the minimum number of fields to impute) can be slightly modified to always identify the failing implicit edits. The failing explicit edits are always known. If the implicit edits are known prior to editing, then general integer-programming methods can exceptionally quickly find the error-localization solution[12] in contrast to the direct integer-programming methods that do not have implicit edits a priori. Furthermore, Winkler [1995] has demonstrated that fast, greedy heuristics – software that often runs at least 10 times faster than general branch-and-bound algorithms – can obtain identical solutions to those obtained by general branch-and-bound algorithms.

If implicit edits are known a priori, then rigorous edit/imputation software can be created that (1) preserves probability distributions, (2) causes all records to satisfy all edits, (3) are fast enough for some of the largest data-mining situations, and (4) can handle enough data to easily deal with the large datasets arising from most sample surveys. This is true in situations involving millions of records.

We next illustrate the concept of an implicit edit with some examples drawn from Fellegi and Holt [1976]. These examples facilitate the discussion of editing software that constitutes Section 7.8.

Example 7.8: Failing at least one edit

Let an edit $E = (Marital\ Status = married) \cap (age = 0 - 14)$ and let r be a data record. Then, $r \in E$ means that record r has failed edit E. In other words, if a record on an individual gives his/her age as being "0–14" and his/her marital status as "married," then either the age or the marital status on the record of that individual must be changed to pass that edit.

We note that if a record fails one or more edits, then at least one field on that failing record must be changed.

Example 7.9: Which field(s) to change?

We summarize three fields on a questionnaire as well as their possible values in Table 7.1.

Suppose that there are two edits:

$$E_1 = (Age = 0 - 14) \cap (Marital\ Status = Ever\ Married),\ and$$

$$E_2 = (Marital\ Status$$

$$= Not\ Now\ Married) \cap (Relationship\ to\ Head\ of\ Household = Spouse)$$

where   "Ever Married" = {Married, Divorced, Widowed, Separated}   and "Not Now Married" = {Single, Divorced, Widowed, Separated}.

Suppose that a record r has the values Age = 0 – 14, Marital Status = Married, and Relationship to Head of Household = Spouse. So, record r fails edit E_1 while it passes edit E_2; symbolically, we write $r \in E_1$ but $r \notin E_2$.

---

[12] See Winkler [1995] or Barcaroli and Venturi [1993 and 1997] for details.

In order to attempt to correct this record, we might consider changing the Marital Status field. It is easy to verify that if we leave both the Age and Relationship to Head of Household fields unchanged, then no Marital Status code produces a record that passes both edits E_1 and E_2. In this example, there is a conflict between the age being no more than 14 years and the relationship to head of household being spouse. Fellegi and Holt say that in this situation a third edit

$$E_3 = (Age = 0 - 14) \cap (Relationship \ to \ Head \ of \ Household = Spouse)$$

is logically implied by edits E_1 and E_2. They call edits of this type *implicit edits*. Fellegi and Holt then state that only after we have identified the implicit edit E_3 are we "in a position to determine systematically the field(s)" within record r "that have to be changed to remove all of its logical inconsistencies."

7.8. Editing Software

Several types of software have been developed to perform automatic editing. Here we focus on six such packages. After brief descriptions of each, we compare them in terms of processing time and usability.

Government agencies have developed these systems for their own purposes and do not make the systems available to external users. The key advantage of the software systems is their adaptability and ease-of-use on different surveys.

The speed of the systems is often determined by the integer-programming algorithms and the generality of edit constraints. Systems that generate implicit edits in a separate procedure prior to production editing are often much faster than systems that do not. The timings of the systems are circa 1999. Newer, faster hardware would yield proportionally faster results.

7.8.1. SPEER

The US Census Bureau has an editing system known as SPEER (Structured Programs for Economic Editing and Referrals) for editing continuous economic

TABLE 7.1. Possible values of three fields on a questionnaire

Field	Possible values
Age	0–14, 15+
Marital status	Single, married, divorced, widowed, separated
Relationship to head of household	Head, spouse of head, other

data that must satisfy ratio edits and a limited form of balancing. The SPEER system has been used at the Census Bureau on several economic surveys since the early 1980s (see Greenberg and Surdi [1984]). Specifically, the SPEER software identifies and corrects elements in data records that must satisfy ratio edits and single-level balancing. By single-level balancing we mean data fields (details and totals) that are allowed to be restricted by at most one balance equation. It is known that ratio edits and balance equations are sufficient tools for editing in more than 99% of economic surveys. Because of the simplicity of the algorithms – there are no general linear inequality edits – SPEER typically processes at least 1,000 records per second even on desktop PCs. The most recent improvements to the SPEER software are described in Draper and Winkler [1997].

7.8.2. DISCRETE

The Census Bureau has an editing system known as DISCRETE for editing discrete data.

Winkler [2002] describes a new algorithm used in DISCRETE that generates implicit edits much faster than previous algorithms. In fact, for very large data situations, Winkler's scheme runs faster – by one to two orders of magnitude – than the system developed by IBM using algorithms of Garfinkel, Kunnathur, and Liepins [1986]. Winkler's scheme has been successfully applied to large data situations that were previously considered intractable.

For situations in which the survey instrument has no skip patterns, Winkler's scheme is thought to generate all implicit edits. When skip patterns are present, it still generates most of the implicit edits. For production editing, DISCRETE typically processes at least 100 records per second.

7.8.3. GEIS/Banff

Statistics Canada (see Kovar, MacMillan, and Whitridge [1991]) has developed a system known as GEIS – Generalized Edit and Imputation System. GEIS uses a series of inequality edits to edit the data of economic surveys. GEIS solves the error localization (EL) problem directly rather than generating implicit edits prior to editing. GEIS has improved computational speed compared to systems developed before GEIS. GEIS typically processes at least 10 records per second. The current version of GEIS is called Banff.

7.8.4. DIESIS

DIESIS (Data Imputation Editing System – Italian Software) is an editing and imputation system developed jointly by ISTAT (the National Statistics Institute of Italy) and the Department of Computer and Systems Science of the University of Roma "La Sapienza." This system is described in Bruni, Reale, and Torelli [2001]. DIESIS implements ideas of Bruni and Sassano [2001], who applied "satisfiability" concepts to discrete data. In this system, implicit edits are generated prior to error localization. Imputation is via a hot deck. DIESIS typically processes at least 100 records per second.

7.8.5. LEO

In the LEO editing system developed at Statistics Netherlands, De Waal [2003b] uses combinations of (1) linear-inequality edits for continuous data and (2) general edits for discrete data. Here, the error localization problem is solved directly by generating failing implicit edits only for the records being processed. Computation is limited by restricting the number of fields that are changed. LEO typically processes at least 10 records per second. At present, LEO is the only system that can process individual edit constraints that simultaneously place constraints on continuous and discrete data.

7.8.6. CANCEIS

CANCEIS (Bankier et al. [1997]) is a system based on somewhat different principles that are still consistent with the principles of a Fellegi–Holt system. It has been used on discrete data for the last three Canadian Censuses and has also been extended to certain types of continuous data. Each edit-failing record is matched against a large number of edit-passing records. Among the closest N records (where N is usually five), a single donor-record is chosen and the edit-failing record is adjusted until it is "near" the donor and no longer fails edits. The advantage of CANCEIS is that it is relatively easy to implement because it does not require set-covering or integer-programming algorithms. It processes more than 100 records per second.

7.8.7. Comparing the Editing Software

Ideally, we would like to compare edit software in terms of ease-of-input of edit constraints, adaptability to different types of survey data, suitability of imputation options for filling in missing or contradictory values of data fields, and speed in day-to-day use. At present, most of the systems described above can deal with the various types of data their agencies encounter. Suitability of imputation has two facets. The first is that the imputed values in fields satisfy edit constraints. Although SPEER and DISCRETE were the first to guarantee that imputed values create records that satisfy edits, all the new versions of the other systems can now do so. The second is that the imputed fields satisfy probabilistic distributional characteristics according to a suitable model. Although the second problem is still open under most conditions, Winkler [2003] connects editing with imputation under the missing-at-random assumption of Little and Rubin [2002]. The key feature of the Winkler procedure [2003] is that all imputations are guaranteed to satisfy edits. If a suitable model for the non-missing-at-random situation is available, then extension to general imputation situations is straightforward. At present, model-building for general imputation is only computationally tractable for small situations (that is, 8–12 variables).

Speed can be an issue on some surveys and most censuses due to the number of records that need to be processed in a limited amount of time. With the increase in computational speed, all systems are able to process 100,000 records in less than 12 hours. Because GEIS and LEO do not generate implicit edits, they solve the error localization problem directly and run relatively slowly. In some situations, the data from a large survey or census can be partitioned into subsets of records that can be processed separately on 10–30 personal computers as is done in the Canadian and UK Censuses. SPEER is able to process the 8 million records of the three US Economic Censuses in less than 3 hours. DISCRETE is able to process the 300 million records of the US Decennial Census on five machines in less than 3 days. DIESIS is at least as fast as DISCRETE.

At present, SPEER, DISCRETE, DIESIS, and a new, faster integer-programming methodology (see Salazar–Gonzalez and Riera–Ledesma [2005]) are the only technologies suitable for applications involving millions of records – for example, cleaning up files that are input into data warehouses for subsequent data mining.

7.9.   Is Automatic Editing Taking Up Too Much Time and Money?

Most editing is designed to clean-up data and put it in a form suitable for producing tabulations. With discrete survey data, most editing is done automatically (using Fellegi–Holt systems or if-then-else rules) with no clerical review. With continuous, economic data, most editing is still done manually. Computerized edit rules may merely identify a moderate proportion of records (say 5–10%) that fail edits and must be corrected manually. Even with Fellegi–Holt systems, subject-matter experts may want to review a sizable portion of the edit-failing records. The key is to reduce the amount of manual editing without seriously affecting the overall quality of the data.

Granquist and Kovar [1997] argue that (1) "well-intentioned individuals edit [survey] data to excruciating detail at tremendous cost" and (2) "more emphasis [must be] placed on using editing to learn about the data collection process, in order to concentrate on preventing errors rather than fixing them." They also say the following:

By far the most costly editing task involves recontacting the respondents in order to resolve questionable answers. The bad-will costs accrued in querying the respondents cannot be neglected, in particular when queries leave the data unchanged. Excessive editing will cause losses in timeliness and hence decrease the data relevance.

The study (see Mustaq and Scheuren [2005]) assessing the cost of collecting State sales tax sales tax referred to in Section 5.4.4 is an example in which the Fellegi–Holt scheme was applied to a small, complex survey with lots of messy data. Although the Fellegi–Holt scheme only changed a small number of data elements, those conducting the study felt that the answer that resulted was reasonable and were also pleased that the work was completed using a relatively small amount of staff time.

7.10. Selective Editing

If the editing delays the publishing/releasing results to the public, then how do the delays affect the use of the data in (1) general circulation publications and (2) research studies of the resulting microdata files? This is an important question, because, although we analysts do not want to over-edit, we want all errors that may be catastrophic for the use of a database to be corrected.

Perhaps the answer[13] is "selective editing" whereby respondent follow-up is limited to queries that have "significant impact." They key idea of selective editing is to target the records having the most effect on certain aggregate estimates. Hidiroglou and Berthelot [1986] and Berthelot and Latouche [1993] have each proposed a scoring method for selecting records for manual editing. These scoring methods can be used instead of a computerized edit/imputation system. De Waal [2003d] and Winkler [1999] argue that the best use of scoring methods is when a set of data has been edited and imputed according to Fellegi– Holt methods that fill-in all of the data in a principled manner. Scoring methods should only be used to target a small number of records for clerical review. Typically, a scoring method (or methods) will target the records having the greatest effect on a few key aggregate estimates. During an open-ended clerical review, some experienced analysts might locate additional anomalies that might not be identified by a fixed set of edit rules as in the Fellegi-Holt model of editing. The new anomalies might be due to changes in a survey form or to alternative methods of data collection (such as via the Internet). It is still an open issue whether extensive clerical review and unsystematic changes by analysts improve the "quality" of the data. This is particularly true when there is no systematic, statistically-sound review process and the analysts' training is limited to ad hoc identification of anomalies.

7.11. Tips on Automatic Editing and Imputation

In data warehouse situations, Fellegi–Holt type editing systems can often be implemented quickly, especially when the developer has extensive knowledge of the data. Frequently, edit tables are all that is required to construct a production edit system. Kovar and Winkler [1996] installed GEIS and SPEER systems for a set of linear-equality edits on continuous economic data in less than a day for each system. In another situation involving discrete data from a small sample

---

[13] This is a suggestion of Granquist and Kovar [1997].

survey, Winkler and Petkunas [1997] developed a production edit system in less than a day.

A single analyst/programmer may be able to create the tables for a Fellegi–Holt system in 1–3 months. For an economic survey with a number of complicated edits, a Fellegi–Holt system would likely require less than 12 hours to run and produce a "corrected" set of data. It is usually considerably more difficult and time-consuming to implement an edit system composed of a consistent set of hard-coded if-then-else rules. Such a system, containing hundreds or thousands of if-then-else rules, might take several programmers 6–12 months to write and might contain serious errors. In extreme situations, 10 analysts might spend 6 months "correcting" edit-failing records and produce "corrected" data files that might not be of sufficiently high quality.

It is also time consuming to build the sophisticated imputation models suggested by Little and Rubin [2002], even if separated from the editing task. At first glance, one might decide to just build an explicit model for the key variables of the survey and use a hot deck for the remaining variables. Unfortunately, it is not clear how to integrate a hot deck with such an explicit model in this situation.

In Chapter 5, we stressed the need to assess the impact of the edit/imputation process. For example, we need to know the number of items imputed and the number and amounts of the items altered by the editing process.

In our experience working with long-tailed distributions, a small number of imputations and/or edits can have substantial impacts on the end results. In such situations, some of us are reluctant to put the edit/imputation process on "automatic pilot"; instead we feel it is crucial to rely on our past experience in closely examining our data. This leads us to the inescapable conclusion that editing and imputation modeling have elements of both art and science, with the art being distinct from the science.

7.12.   Where Are We Now?

We have now completed the first seven chapters of the book. While we have discussed the need to prevent errors from entering our lists/databases, the emphasis thus far has been on detecting and repairing errors. This is reasonable as our intended audience is an analyst operating within a large organization. The analyst could bring such data problems to those above her in the organization's chain of command; this might even do some good, eventually. However, for us, it has the feel of praying for rain – too passive for our taste. On the other hand, the methods relating to record linkage that we describe in Chapters 8–13 can directly improve fitness for use. Using record linkage techniques, the analyst can add real value to her data, often at modest cost.

8
Record Linkage – Methodology

8.1. Introduction

Record linkage, in the present context, is simply the bringing together of information from two records that are believed to relate to the same entity – for example, the same individual, the same family, or the same business. This may entail the linking of records within a single computer file to identify duplicate records. A few examples of this were mentioned in Chapter 3. Alternatively, record linkage may entail the linking of records across two or more files. The challenge lies in bringing together the records for the same individual entities. Such a linkage is known as an *exact match*. This task is easiest when unique, verified identification numbers (e.g., Social Security Numbers) are readily available. The task is more challenging when (1) the files lack unique identification numbers, (2) information is recorded in non-standardized formats, and (3) the files are large. In this case, names, addresses, and/or dates of birth are frequently used in the matching process. A large number of additional examples of record linkage are described in Part Three. Most of these involve linking records across multiple files.

This chapter begins with a brief historical section. We then consider two basic types of record linkage strategies: *deterministic* and *probabilistic*, both of which are considered to be a type of exact matching.[1] Chapter 9 contains

---

[1] An alternative approach – not considered further in this work – is known as *statistical matching*. A *statistical match* is defined as a match in which the linkage of data for the same unit across files either is not sought or is sought but finding such linkages is not essential to the approach. In a statistical match, the linkage of data for similar units rather than for the same unit is acceptable and expected. Statistical matching is ordinarily employed when the files being matched are probability samples with few or no units in common; thus, linkage for the same unit is not possible for most units. Statistical matches are made based on similar characteristics, rather than on unique identifying information, as in the usual exact match. Other terms that have been used for statistical matching include "synthetic," "stochastic," "attribute," and "data" matching. The reader should see Office of Federal Statistical Policy and Standards [1980] for more details on statistical matching.

an extended discussion of parameter estimation for the probabilistic model that is the main focus of this chapter – the Fellegi–Sunter Record Linkage Model. Chapters 10–13 describe tools for linking records by name and/or address.

8.2.   Why Did Analysts Begin Linking Records?

Record linkage techniques were initially introduced about 50 years ago. Fellegi [1997] suggests that this resulted from the confluence of four developments:

- First, the post-war evolution of the welfare state and taxation system resulted in the development of large files about individual citizens and businesses.
- Second, new computer technology facilitated the maintenance of these files, the practically unlimited integration of additional information, and the extraction of hitherto unimaginably complex information from them.
- Third, the very large expansion of the role of government resulted in an unprecedented increase in the demand for detailed information which, it was thought, could at least partially be satisfied by information derived from administrative files which came about precisely because of this increase in the role of the government.
- But there was a fourth factor present in many countries, one with perhaps the largest impact on subsequent developments: a high level of public concern that the other three developments represented a major threat to individual privacy and that this threat had to be dealt with.

8.3.   Deterministic Record Linkage

In deterministic record linkage, a pair of records is said to be a *link* if the two records agree exactly on each element within a collection of identifiers called the *match key*. For example, when comparing two records on last name, street name, year of birth, and street number, the pair of records is deemed to be a link only if the names agree on all characters, the years of birth are the same, and the street numbers are identical.

Example 8.1: Linking records when the identification schemes are different

The Low-Income Housing Tax Credit (LIHTC) Program in the United States provides tax credits for the acquisition, rehabilitation, or new construction of multifamily rental housing (e.g., apartment buildings) targeted to lower-income households. Some researchers think this program is the most important means of creating affordable housing in the United States today. The US Department of Housing and Urban Development (HUD) maintains a database on over 22,000 projects that have participated in the LIHTC program since its inception in 1986. In addition, HUD maintains a database of multifamily apartment projects whose mortgages have been insured by the Federal Housing Administration (FHA).

Analysts working on behalf of FHA wanted to establish that a number of apartment projects with FHA-insured mortgages received tax credits under the LIHTC program. The portion of the HUD multifamily database used consisted of almost 7,000 mortgages originated between 1986 and 1998. Unfortunately, the two databases employed incompatible identification schemes, so the identification numbers could not be used to link records between the two databases. A deterministic (i.e., exact) matching scheme was implemented that involved (1) the five-digit Zip Code of the address of the apartment project and (2) the first four characters of the name of the apartment project. Here the match key was a string of nine characters for each project (five from the Zip Code and four from the name) and the exact matching scheme was applied to such strings. This scheme identified approximately 300 projects that were receiving such tax credits and whose mortgages were FHA-insured.

Although this approach was naïve and undoubtedly missed many matches, it nonetheless met the needs of the analysts who wanted this work done. It is likely that more than 300 of the 7,000 records should have been linked. If the analysts wanted to make an estimate of the proportion of FHA-insured apartments that had received tax credits under the LIHTC program, then 0.042 (300/7000) is likely to be too low. To determine a greater number of links, it might have been necessary to weaken the criterion for exact agreement on Zip Code to agreement on the first three characters of Zip Code (i.e., generally the same metropolitan area). With slightly greater sophistication, it might be possible to find all pairs of records from the two files that agreed on two four-character substrings of the name fields. In each situation, we would still need to use other characteristics of the pairs of records to determine whether a pair should actually be designated as a match. In each of the situations where we match pairs of records on the weakened characteristics, we try to increase the number of matches but, as an undesired consequence, increase the probability of false matches as well.

8.4. Probabilistic Record Linkage – A Frequentist Perspective

Fellegi and Sunter [1969] mathematically formalized probabilistic record linkage based on earlier work of Newcombe et al. [1959]. Under the Fellegi–Sunter model, pairs of records are classified as links, possible links, or non-links. We consider the product space, $A \times B$, of two files we denote by A and B.

The first step in the record linkage procedure is to edit all fields of the records in files A and B into a standardized format. The next step is to compare the records between the two files.

To do this, we consider $A \times B$ to be partitioned into two sets: the set of true matches, denoted by M, and the set of non-matches, denoted by U.

Probabilistic record linkage assigns a numerical value to (the similarity of) a pair of records, r, as a monotonically increasing function of the ratio, R, (e.g., $\ln(R)$), of two conditional probabilities:

$$R = \frac{P(\gamma \in \Gamma \mid r \in M)}{P(\gamma \in \Gamma \mid r \in U)} \tag{8.1}$$

where γ is an arbitrary agreement pattern in a comparison space Γ.

Example 8.2: Likelihood ratio under Fellegi–Sunter model

Let Γ consist of the eight $(= 2^3)$ possible patterns representing simple agreement or disagreement on three fields (e.g., the last name, street name, and street number) of a pair of records, r, drawn from $A \times B$. Then, we can rewrite Equation (8.1) as

$$R = \frac{P(\textit{agree on last name, agree on street name, agree on street number} \mid r \in M)}{P(\textit{agree on last name, agree on street name, agree on street number} \mid r \in U)}$$

when the pair of records, r, is observed to be in exact agreement on all three fields. If instead the pair of records, r, disagrees on the street number, then we would have

$$R = \frac{P(\textit{agree on last name, agree on street name, disagree on street number} \mid r \in M)}{P(\textit{agree on last name, agree on street name, disagree on street number} \mid r \in U)}$$

All record pairs that agree on last name, street name, and street number might be assigned equal probabilities. Alternatively, the probabilities might depend on the specific values of the fields. For instance, pairs in which both last names were "Zabrinsky" might be assigned a higher probability than pairs in which both last names were "Smith." We discuss various techniques for computing these probabilities in Chapter 9.

The ratio of Expression (8.1) is large for agreement patterns that are found frequently among matches but are rarely found among non-matches. The ratio is small for patterns more consistent with non-matches than matches. The ratio can be viewed (using the terminology of mathematical statistics) as a likelihood ratio. The problem of choosing one status from link, possible link, or non-link can also be viewed as a double hypothesis-testing problem. In addition, it can be viewed as a classification problem in statistics or a machine learning problem (learning the concept of a match) in computer science.

8.4.1. Fellegi–Sunter Decision Rule

Fellegi and Sunter [1969] proposed the following decision rule in such cases:

If $R \geq Upper$, then call the pair, r, a *designated match* or a *designated link*.

If *Lower* < *R* < *Upper*, then call the pair, *r*, a *designated potential match* or a *designated potential link* and assign the pair to clerical review. (8.2)
If *R* ≤ *Lower*, then call the pair, *r*, a *designated non-match* or a *designated non-link*.

Fellegi and Sunter [1969] showed that this decision rule is optimal in the sense that for any pair of fixed bounds on the error rates on the first and third regions of R, the middle region is minimized over all decision rules on the same comparison space Γ. The cutoff thresholds *Upper* and *Lower* are determined by the a priori error bounds on false matches and false non-matches. Rule (8.2) agrees with intuition. If $\gamma \in \Gamma$ consists primarily of agreements, then it is intuitive that $\gamma \in \Gamma$ would be more likely to occur among matches than non-matches and the ratio (8.1) would be large. On the other hand, if $\gamma \in \Gamma$ consists primarily of disagreements, then the ratio (8.1) would be small. In actual applications, the optimality of the decision rule (8.2) is heavily dependent on the accuracy of the estimates of the probabilities of (8.1).

In Figure 8.1, we plot data obtained from the1988 dress rehearsal of the 1990 US decennial census. The weight (rounded to the nearest .1) on the horizontal axis is the natural logarithm of the likelihood ratio given in equation (8.1). The values on the vertical axis are the natural logarithms of 1 plus the number of pairs having a given weight. We plot the data separately for matches (identified by "*") and non-matches (identified by "Þ"). The pairs between the upper and lower cutoffs (vertical bars) generally consist of individuals in the same household (i.e., agreeing on address) and missing both first name and age. The true match status was determined after clerical review, field follow-up, and two rounds of adjudication.

8.4.2. Conditional Independence

Sometimes we can rewrite the probabilities above as

$P(agree\ on\ last\ name,\ agree\ on\ street\ name,$

$agree\ on\ street\ number \mid r \in M) = P(agree\ on\ last\ name \mid r \in M)$

$P(agree\ on\ street\ name \mid r \in M)P(agree\ on\ street\ number \mid r \in M)$

and

$P(agree\ on\ last\ name,\ agree\ on\ street\ name,$

$disagree\ on\ street\ number \mid r \in M) = P(agree\ on\ last\ name \mid r \in M)$

$P(agree\ on\ street\ name \mid r \in M)P(disagree\ on\ street\ number \mid r \in M)$

Such equations are valid under the assumption of *conditional independence*. We assume that conditional independence holds for all combinations of fields (variables) that we use in matching.

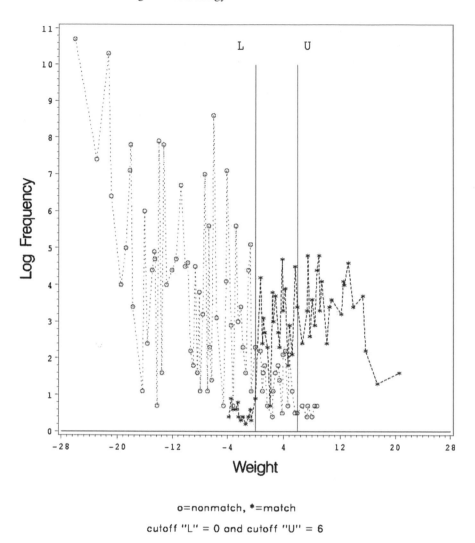

o=nonmatch, *=match

cutoff "L" = 0 and cutoff "U" = 6

FIGURE 8.1. Log Frequency vs Weight Matches and Nonmatches Combined.

The probabilities P(agree first | M), P(agree last | M), P(agree age | M), P(agree first | U), P(agree last | U), and P(agree age | U) are called *marginal probabilities*. The marginal probabilities $P(\cdot|M)$ and $P(\cdot|U)$ are called m- and u-probabilities, respectively. The base 2 logarithm of the ratio of the probabilities, $\log_2(R)$, is called the *matching weight*, *total agreement weight*, *binit weight* or *score*. The logarithms of the ratios of probabilities associated with individual fields are called the *individual agreement weights*. The m- and u-probabilities are also referred to as *matching parameters*. The main advantage of the conditional independence assumption is that it makes it relatively straightforward to estimate the m- and u-probabilities. Even in straightforward situations, estimation of the m- and u-probabilities requires practical experience in statistical analysis particularly when dealing with heterogeneous individual agreement weights.

8.4.3. Validating the Assumption of Conditional Independence

Although in theory the decision rules of Fellegi and Sunter hold for general classes of distributions on the matches, M, and the non-matches, U, Fellegi and Sunter provide a number of straightforward computational procedures that are only valid under a conditional independence assumption (see Fellegi and Sunter [1969; pp. 1189–1190]). As they indicate, their conditional independence assumption may not be so crucial in practice. Parameters (probabilities) estimated when the conditional independence assumption is violated may still yield accurate decision rules in many situations. Winkler [1993, 1994] demonstrated that good decision rules were possible with the 1990 Decennial Census data even though Thibaudeau [1989] and Winkler [1989a] showed that such data could have substantial departures from conditional independence.

The failure of the conditional independence assumption may be easy to spot. Two files that have 1,000 records each yield one million pairs. In Newcombe's pioneering work, only pairs were considered that agreed on a field such as the surname. In more recent times, a geographical identifier such as the nine-digit Zip Code in the United States (representing about 70 households) might be used. If a pair of records agrees on surname, then it is more likely to agree on other characteristics such as date of birth. If a pair agrees on the nine-digit Zip Code, then it is more likely to simultaneously agree on characteristics such as last name, house number, and street name. This is true regardless of whether the pairs are matches, M, or non-matches, U. Although such departures from conditional independence can be quite pronounced (i.e., the joint probabilities may be appreciably different from the corresponding products of marginal probabilities based on individual fields), the decision rules can still be quite accurate.

8.4.4. Weighting

The *individual agreement weight, w_i,* of the ith field of the rth pair of records is computed from the m- and u- probabilities as follows. Using the notation

$$m_i = \text{Pr ob[agreement in field i } | \text{ r} \in M]$$

and

$$u_i = \text{Pr ob[agreement in field i } | \text{ r} \in U],$$

we obtain

$$w_i = \begin{cases} \log_2 \left(\frac{m_i}{u_i} \right) & \text{if agreement in field i} \\ \log_2 \left(\frac{(1-m_i)}{(1-u_i)} \right) & \text{if otherwise.} \end{cases}$$

We further assume here that we want to base our scores on the entries in n fields of each record. In this case, we want to consider the n-long vector $\gamma = (\gamma_1, \ldots, \gamma_n)$ and the n-dimensional cross-product space $\Gamma = \Gamma_1 \times \cdots \times \Gamma_n$. This allows us to write the ratio, R, of the conditional probabilities as

$$R = \frac{P[(\gamma_1, \ldots, \gamma_n) \in \Gamma_1 \times \cdots \times \Gamma_n \mid r \in M]}{P[(\gamma_1, \ldots, \gamma_n) \in \Gamma_1 \times \cdots \times \Gamma_n \mid r \in U]}$$

so that

$$\log_2(R) = \log_2 \left\{ \frac{P[(\gamma_1, \ldots, \gamma_n) \in \Gamma_1 \times \cdots \times \Gamma_n \mid r \in M]}{P[(\gamma_1, \ldots, \gamma_n) \in \Gamma_1 \times \cdots \times \Gamma_n \mid r \in U]} \right\}.$$

Now we can use our conditional independence assumption to rewrite the last equation as

$$\log_2(R) = \log_2 \left\{ \prod_{i=1}^{n} \frac{P[\gamma_i \in \Gamma_i \mid r \in M]}{P[\gamma_i \in \Gamma_i \mid r \in U]} \right\} = \sum_{i=1}^{n} \log_2 \left\{ \frac{P[\gamma_i \in \Gamma_i \mid r \in M]}{P[\gamma_i \in \Gamma_i \mid r \in U]} \right\}.$$

Moreover, each term of the sum in the last equation is just w_i. Hence, the matching weight or score of pair r is simply the sum of the weights

$$\log_2(R) = \sum_{i=1}^{n} w_i.$$

Large positive matching weights (or scores) suggest strongly that the pair of records is a link; while large negative scores suggest non-links. Although any one piece of information might not be sufficient to distinguish links from non-links, in most cases the ensemble of the information (i.e., the matching weight) creates sufficient evidence for the computer to decide.

8.4.5. Typographical Errors/Variations

In practical situations, names and other fields used in the matching process may contain typographical error. In this work, to be precise, we define a *typographical error* as a variation in the representation or spelling of an entry in a field. Such typographical error, for example, may cause the surname of an individual entered on two separate records to fail to agree on an exact letter-by-letter basis. In each of the situations depicted in Example 8.3 except the first, we have a typographical variation in the given name of an individual.

Example 8.3: Typographical Variation of a Given Name

- David – given name, likely name on birth certificate
- Dave – common nickname of individual
- Daue – typographical error, possibly due to keying of handwritten form
- "b" – blank or missing value

- Edward – individual usually uses middle name.

Example 8.4: Disagreeing Pairs of Given Names

1. (David, Dave) – actual given name versus nickname
2. (Dave, David) – actual given name versus nickname
3. (Dave, Daue) – typographical variation in both entries of given name.

If a name was a straightforward name such as 'David Edward Smith' and the name was always represented without error in computer files, then we could more easily perform matching. If there were no typographical error and we could always easily bring together pairs, then among matches, M, the agreement probabilities would all be 1. For example, we would have $P[\text{agree given name} \mid M] = 1$.

In matching situations, we need to be able to estimate each of the probabilities associated with the agreement (and disagreement) on individual fields. Given representative training data or some of the more advanced parameter estimation methods that we discuss in Chapter 9, we often have situations where $P[\text{agree given name} \mid M] \leq .92$. We can interpret the probability, $P[\text{agree given name} \mid M]$, as the average agreement probability on first names among matches. From the typographical error viewpoint, we can interpret $P[\text{agree given name} \mid M] = .92$ as meaning that an average pair among matches has a 0.08 probability of disagreeing on given name due to typographical variation in the given name field.

Examples 8.3 and 8.4 raise the question as to whether we can use look-up tables to correct for nicknames and spelling variation/error as is done with many commonly occurring words in word processing software. The answer is: "only partially and not reliably." For example "Bobbie" could be

- A woman's given name (i.e., on a birth certificate),
- A nickname for a man whose given name is "Robert," or
- A nickname for a woman whose given name is "Roberta".

Moreover, we can't "clean-up" many foreign (i.e., non-typically English or American) given names or surnames. It is also not a good idea to "clean-up" many other fields such as house number.

8.4.6. Calculating Matching Weights

Examples 8.5 and 8.6 illustrate the calculation of m-probabilities, u-probabilities, agreement weights, disagreement weights, and matching weights.

Example 8.5: Gender errors

Consider two databases that are to be matched. In each database the value of the "gender" field is incorrect 10% of the time on each member of the pairs that are matches. This means that there is a 10% probability of typographical error in the gender values from the first file as well as the second file. We also assume that the gender value is never blank. Compute the m-probability for this field.

Solution

The probability that both gender field values are correct is $(.9) \cdot (.9) = .81$. The probability that both gender field values are incorrect is $(.1) \cdot (.1) = .01$. Therefore, the desired m-probability is $.81 + .01 = .82$.

More generally, for fields that take non-binary values, if both values are incorrect, then the probability of agreement, given a match, may be closer to 0.0 than to 0.01. Obvious examples are first names and last names that can take many values.

Example 8.6: Calculating a Matching Weight

We wish to compare two records based on the values of two of their fields: gender and Social Security Number. Assume, as calculated above, that the gender fields have (1) an m-probability of .82 and (2) a u-probability of .5 because there are an equal number of men and women. Assume further that (1) the m-probability for the Social Security Number is .6 and (2) the u-probability for the Social Security Number is 10^{-7} – one in 10 million. Compute both the individual agreement weights and disagreement weights for the two variables. Also, compute the matching weight if the two records are observed to have identical Social Security Numbers but the values of their gender variables disagree. Assume conditional independence.

Solution

From Section 8.4.4, we have the agreement and disagreement weights for the Social Security Number field as respectively

$$\log_2 \left(\frac{m}{u} \right) = \log_2 \left(\frac{.6}{10^{-7}} \right) = \log_2 \left(6 \cdot 10^6 \right)$$

and

$$\log_2 \left(\frac{1-m}{1-u} \right) = \log_2 \left(\frac{1-.6}{1-10^{-7}} \right) = \log_2 \left(\frac{.4}{.9999999} \right).$$

The agreement and disagreement weights for the gender field are respectively

$$\log_2 \left(\frac{m}{u} \right) = \log_2 \left(\frac{.82}{.5} \right) = \log_2 (1.64)$$

and

$$\log_2\left(\frac{1-m}{1-u}\right) = \log_2\left(\frac{1-.82}{1-.5}\right) = \log_2\left(\frac{.18}{.5}\right) = \log_2(.36).$$

The matching weight or score for this example is

$$\log_2(R) = \sum_{i=1}^{2} w_i = \log_2\left(6 \cdot 10^6\right) + \log_2(.36) = \log_2\left((2.16) \cdot 10^6\right).$$

In Example 8.6, we made the assumption that the m-probability for the Social Security Number field is 0.6. In situations where both lists have "verified" Social Security Numbers, we expect the probability to be very high (e.g., 0.995). If one of the files being matched has self-reported Social Security Numbers (which are often blank or in error), then the m-probability for the Social Security Number field may be substantially less than 0.4.

8.5.   Probabilistic Record Linkage – A Bayesian Perspective

Most Bayesian statisticians would feel more comfortable considering a probability of the form of

$$P[r \in M \mid \gamma \in \Gamma]. \tag{8.3}$$

Here, we are considering the probability of a match given the observed data. This is the form Bayesians prefer to that used in the numerator of Expression (8.1). It turns out that it is apparently *not* a particularly simple matter, in most practical record linkage situations, to estimate probabilities of the form of Expression (8.3).

Nonetheless, Belin and Rubin [1995] do indeed present a scheme for computing probabilities of the form of Expression (8.3). In order to model the shape of the curves of matches and non-matches, they require the true matching status for a representative set of pairs. They postulate that the distribution of weights of all of the pairs is a mixture of two distributions. They propose that the analyst first transform the data (i.e., the weights) using a Box-Cox transformation of the form

$$\Psi(w_i; \gamma; \omega) = \begin{cases} \frac{w_i^\gamma - 1}{\gamma \omega^{\gamma-1}} & if \ \gamma \neq 0 \\ \omega \ln(w_i) & if \ \gamma = 0 \end{cases}$$

where ω is the geometric mean of the weights $w_1, w_2, \ldots, w_n$. This is intended to convert the weights to a mixture of two transformed normal distributions. By "transformed normal distribution," Belin and Rubin "mean that for some unknown values of γ and ω, the transformed observations $\Psi(w_i; \gamma, \omega)$ are normally distributed."

They then propose employing a scheme known as the EM algorithm to obtain maximum likelihood estimates of the unknown parameters of the two components of the normal mixture model. Finally, they suggest using the SEM algorithm (see Meng and Rubin [1991]) to obtain estimates of the standard errors of the parameters fit using the EM algorithm.

Belin and Rubin [1995] contend that the frequentist "methods for estimating the false-match rates are extremely inaccurate, typically grossly optimistic."

On the other hand, Winkler [2004] states that the methods of Belin and Rubin [1995] only work well in a narrow range of situations where (1) the curves associated with the matches, M, and the non-matches, U, are somewhat separated, (2) each curve has a single mode, and (3) the failure of the conditional independence assumption is not too severe. The Belin and Rubin method is thought to work particularly well when the curves are obtained via a 1–1 matching rule. However, with many administrative lists, business lists, and agriculture lists, the curves are not well-separated and the business list weighting curves typically do not have a single mode; consequently, the methods of Belin and Rubin are not appropriate for such applications. The weighting curves associated with the matches can also be multi-modal if

- there is name or address standardization failure,
- there are certain types of typographical error,
- one of the source lists is created by merging several different lists, or
- the data entry clerks either have heterogeneous skill levels or learn to override some of the edits in the software.

8.6. Where Are We Now?

In this chapter we provided some historical background; gave the main decision rule for classifying pairs into matches, clerical pairs, and non-matches; gave examples of how the probabilities in the decision rule are computed under a conditional independence assumption; and indicated how typographical error can make matching more difficult. In the following chapters, we will more fully explore the estimation of matching parameters (i.e., probabilities) in practice, the use of parsing and standardization to help prepare files for parameter estimation and subsequent matching, the use of "phonetically" encoded fields to attempt to overcome many typographical errors, the use of blocking methods to bring together relatively small subsets of pairs in situations where there are large numbers of pairs in the product space of the two files A and B, and methods of automatically dealing with typographical errors (primarily very minor ones) among pairs that are brought together.

9
Estimating the Parameters of the Fellegi–Sunter Record Linkage Model

In this chapter, we discuss several schemes for estimating the parameters (i.e., the m-and u-probabilities) of the Fellegi–Sunter model discussed in Chapter 8.

9.1. Basic Estimation of Parameters Under Simple Agreement/Disagreement Patterns

For each $\gamma \in \Gamma$, Fellegi and Sunter [1969] considered

$$P(\gamma) = P(\gamma \mid r \in M)P(r \in M) + P(\gamma \mid r \in U)P(r \in U)$$

and noted that the proportion of record pairs, r, having each possible agreement/disagreement pattern, $\gamma \in \Gamma$, could be computed directly from the available data. For example, if $\gamma = (\gamma_1, \gamma_2, \gamma_3)$ consists of a simple agree/disagree (zero/one) pattern associated with three variables, then a typical value for γ would be $(1, 1, 0)$. Then, by our usual conditional independence assumption, there exist vector constants (marginal probabilities) $m = (m_1, m_2, \ldots, m_n)$ and $u = (u_1, u_2, \ldots, u_n)$ such that, for all 2^n possible values of $(\gamma_1, \gamma_2, \ldots, \gamma_n)$

$$P[(\gamma_1, \gamma_2, \ldots, \gamma_n) \mid r \in M] = \prod_{i=1}^{n} m_i^{\gamma_i} (1 - m_i)^{1-\gamma_i}$$

and

$$P[(\gamma_1, \gamma_2, \ldots, \gamma_n) \mid r \in U] = \prod_{i=1}^{n} u_i^{\gamma_i} (1 - u_i)^{1-\gamma_i}$$

For the case in which $n \geq 3$, Fellegi and Sunter [1969] showed how to use the equations above to find solutions for the $2n + 1$ independent parameters – $m_1, m_2, \ldots, m_n, u_1, u_2, \ldots, u_n$, and $P[M]$. (We obtain $P[U]$ as $P[U] = 1 - P[M]$.) The reader can obtain further details from Fellegi and Sunter [1969].

9.2.   Parameter Estimates Obtained via Frequency-Based Matching[1]

If the distribution of the attribute values for a field is not uniform, then a value-specific (frequency-based), or outcome-specific, weight can be introduced. Frequency-based weights are useful because they can account for the fact that a specific pair of surnames such as (Zabrinsky, Zabrinsky) occurs less often in the United States than a pair of surnames such as (Smith, Smith). This is useful because names such as Smith and Jones that are common in the general population of the United States may not be as effective as a relatively rare name such as Zabrinsky in distinguishing matches.

Moreover, surnames such as Martinez and Garcia have more distinguishing power in Minneapolis than in Los Angeles because Hispanic surnames are so much more common in Los Angeles.

How can such phenomena be incorporated into our Fellegi–Sunter models? The answer is by modifying the model to give a larger agreement weight to infrequently occurring surnames. This can be justified further by noting that if γ represents the surnames on a pair of records, r, then

$$P(r \in M \mid agreement\ on\ Zabrinsky) > P(r \in M \mid agreement\ on\ Smith)$$

which implies that

$$P(r \in U \mid agreement\ on\ Smith) > P(r \in U \mid agreement\ on\ Zabrinsky).$$

Using the last two inequalities in conjunction with Bayes' Theorem, we can show that

$$\frac{P(agreement\ on\ Zabrinsky \mid r \in M)}{P(agreement\ on\ Zabrinsky \mid r \in U)} > \frac{P(agreement\ on\ Smith \mid r \in M)}{P(agreement\ on\ Smith \mid r \in U)}.$$

Fellegi and Sunter [1969, pp. 1192–1194] propose a solution to this problem. To simplify their ideas, we begin by assuming that neither file A nor file B contains any duplicate records. We also assume that the true frequencies of specific fields (e.g., the surname of an individual) in files A and B, respectively, are

$$f_1, f_2, \ldots, f_m$$

where the number of records in file A is $N_A = \sum_{j=1}^{m} f_j$, and

$$g_1, g_2, \ldots, g_m$$

---

[1] For further details on this topic, the interested reader should see Section 3.3.1 of Fellegi and Sunter [1969] or Winkler [1989b].

where the number of records in file B is $N_B = \sum_{j=1}^{m} g_j$. If the mth surname or string, say "Smith", occurs f_m times in file A and g_m times in file B, then "Smith" occurs $f_m g_m$ times in the pairs of records that constitute $A \times B$. The corresponding true frequencies in the set of matching pairs, M, are similarly assumed to be

$$h_1, h_2, \ldots, h_m$$

where the number of records in file M is $N_M = \sum_{j=1}^{m} h_j$. We note that for $j = 1, 2, \ldots, m$, we have $h_j \leq \min \left(f_j, g_j \right)$. For some applications, we assume that

- $h_j = \min \left(f_j, g_j \right)$,
- $P[agreement\ on\ string\ j \mid r \in M] = \frac{h_j}{N_M}$, and
- $P[agreement\ on\ string\ j \mid r \in U] = \frac{f_j \cdot g_j - h_j}{N_A \cdot N_B - N_M}$.

In practice, we must use observed or synthetic values in the actual files being matched because we have no way of knowing the true values. Practitioners have used a variety of schemes to construct the frequencies, h_j. These depend upon how typographical errors are modeled and what simplifying assumptions are made. (For example, a surname such as "Smith" might be recorded as "Snith" due to a typographical error.) The typographical errors then distort the frequency counts in the observed files. Fellegi and Holt [1969] presented a method for dealing with some typographical errors.

In order to obtain numerically stable estimates of the simple agree/disagree probabilities, we use the EM algorithm

Thibaudeau [1989] and Winkler [1989c, 1992] have used the EM algorithm in a variety of record linkage situations. In each, it converged rapidly to unique limiting solutions over different starting points. The major difficulty with the EM algorithm or similar procedures is that it may produce solutions that partition $A \times B$ into two sets that differ substantially from the desired sets of true matches, M, and true non-matches, U. We describe the specifics of the EM algorithm in Section 9.4, later in this chapter. Winkler [1989b] provided a slight generalization of the Fellegi–Sunter method for frequency-based matching. Specifically, Winkler showed that a "surrogate" typographical error rate could be computed using the EM algorithm and that the frequency-based weights could be "scaled" to the EM-based weights. The reader should see Winkler's paper for more details.

9.2.1. Data from Prior Studies

In some applications, the frequency tables are created "on-the-fly" – i.e., from the data of the current application. Alternatively, we could create them using a large reference file of names constructed from one or more previous studies. The advantage of "on-the-fly" tables is that they can use different relative frequencies in different geographic regions – for instance, Hispanic surnames in Los Angeles,

Houston, or Miami and French surnames in Montreal. The disadvantage of "on-the-fly" tables is that they must be based on data files that cover a large percentage of the target population. If the data files contain samples from a population, then the frequency weights should reflect the appropriate population frequencies. For instance, if two lists of companies in a city are used and "George Jones, Inc." occurs once on each list, then a pair should not be designated as a match using name information only. Corroborating information, such as the business's address, should also be used because the name "George Jones, Inc." may not uniquely identify the establishment. Moreover, the population should be the population of interest to the list-builder; for example, if The National Council of La Raza is building a list of registered Hispanic voters in California, then the frequencies should represent the Hispanic subpopulation of California rather than the frequencies in the general population.

This use of frequency tables in conjunction with the Fellegi–Sunter model has often worked well in practice, especially when the typographical error rate is constant over all of the possible values of the surname. Moreover, we note that frequency-based matching could be applied to other fields in the database besides surname. While this value-specific approach can be used for any matching element, strong assumptions must be made about the independence between agreements on specific values of one element versus agreement on other fields.

Three other circumstances can adversely affect this frequency-based matching modification:

- The two files A and B are samples.
- The two files A and B do not have much overlap.
- The corresponding fields in files A and B have high rates of typographical error.

Example 9.1: Matching marriage and birth records in British Columbia

When matching records on marriages and births that occurred in British Columbia during 1946–1955, Newcombe et al. [1959] considered as one factor the frequencies of the husbands' and brides' last names. A positive value (or weight) was added to the score if the names were in agreement in the two databases. The value added depended on how rare or common the last names were in the file as a whole. Other data fields, such as age and place of birth, were treated similarly.

9.3. Parameter Estimates Obtained Using Data from Current Files

Before discussing the EM algorithm, we describe an alternative scheme for computing the u- and m-probabilities "on-the-fly."

9.3.1. u-Probabilities

Assume that File A has 1,000 records and that File B also has 1,000 records. Then, at most 1,000 of the 1,000,000 record pairs in $A \times B = M \cup U$ can be matches. This suggests that the u-probabilities can be reasonably approximated by random agreement weights. We can approximate P [*agree on first name* $| r \in U$], for example, by counting the number of pairs in $A \times B$ that agree on the first name and dividing the result by the number of pairs in $A \times B$ because almost all of the pairs in $A \times B$ are in U.

Alternatively, we may only look at certain subsets of $A \times B$. For example, we might only consider pairs whose Zip Codes are identical. This approach, known as *blocking*[2], usually leads to a substantial reduction in the number of pairs that need to be considered. When blocking is used, the u-probabilities can be computed as

P [*agree on first name* $| r \in U$ *and there is agreement on the blockings criteria*].

Fellegi and Sunter [1969] provide details on the adjustments that need to be made to the matching rules when blocking is used.

9.3.2. m-Probabilities

Assume that we have used the initial guess of m-probabilities given in the previous section and performed matching. We could draw a sample of pairs, perform follow-up (manual review, etc.) to determine true match status, and re-estimate the m-probabilities based on the sample for which we know the truth. Newcombe [1959] suggested this type of iterative-refinement procedure. It typically yields good estimates of the m-probabilities that work well in the matching decision rules. Winkler [1990] compared the m-probabilities obtained via the iterative-refinement procedure with m-probabilities obtained via several other methods. Surprisingly, with certain high-quality files, EM-based estimates of m-probabilities that do not require any training data or follow-up out-performed the m-probabilities obtained via the iterative refinement procedure.

The iterative-refinement methodology for estimating the m- and u-probabilities is known to work well in practice and is used in several commercial software packages. It can work well with messy, inconsistent data. This type of data is typically encountered in business lists, agriculture lists, and some administrative lists.

9.4.   Parameter Estimates Obtained via the EM Algorithm

We begin by assuming that the cross-product space $A \times B = M \cup U$ consists of N pairs of records where each record has n fields. We define

$$\gamma_i^j = \begin{cases} 1 & \textit{if field i is identical on both of the records of record pair } r_j \\ 0 & \textit{otherwise} \end{cases}$$

---

[2] Blocking is the subject of Chapter 12.

where $i = 1, 2, \ldots, n$ and $j = 1, 2, \ldots, N$. We also define

$$\gamma^j = \left\{ \gamma_1^j, \gamma_2^j, \ldots, \gamma_n^j \right\}$$

and

$$\gamma = \left\{ \gamma^1, \gamma^2, \ldots, \gamma^N \right\}.$$

Furthermore, we define the components of the m- and u-probabilities

$$m = \{m_1, m_2, \ldots, m_n\} \text{ and } u = \{u_1, u_2, \ldots, u_n\}$$

as

$$m_i = P\left[\gamma_i^j = 1 \mid r_j \in M\right] \text{ and } u_i = P\lfloor\gamma_i^j = 1 \mid r_j \in U\rfloor$$

for a randomly selected pair of records r_j.

We next define p as the proportion of record pairs that match:

$$p = \frac{Number\ of\ Pairs\ of\ Records\ in\ set\ M}{N}.$$

The N record pairs of interest are distributed according to a finite mixture model with the unknown parameters $\Phi = (m, u, p)$. The goal is to use an EM algorithm to estimate these unknown parameters, especially the vector m.

We next consider the complete data vector $\langle g, \gamma \rangle$ where $g = (g_1, g_2, \ldots, g_N)$ is a vector of indicator functions with

$$g_j = \begin{cases} 1 & \text{if } r_j \in M \\ 0 & \text{if } r_j \in U. \end{cases}$$

Then the complete data likelihood function is

$$f(g, \gamma \mid m, u, p) = \prod_{j=1}^{N} \left(p \cdot P\left[\gamma^j \mid r_j \in M\right]\right)^{g_j} \left((1-p)P\left[\gamma^j \mid r_j \in U\right]\right)^{1-g_j},$$

and the complete data log-likelihood is

$$\ln\left(f(g, \gamma \mid m, u, p)\right) = \sum_{j=1}^{N} g_j \cdot \ln\left(p \cdot P[\gamma^j \mid r_j \in M]\right) + \sum_{j=1}^{N} (1-g_j) \cdot \ln\left((1-p) \cdot P[\gamma^j \mid r_j \in U]\right).$$

We assume (e.g., by invoking the conditional independence assumption) that

$$P[\gamma^j \mid r_j \in M] = \prod_{i=1}^{n} m_i^{\gamma_i^j}(1 - m_i)^{1-\gamma_i^j}$$

and

$$P[\gamma^j \mid r_j \in U] = \prod_{i=1}^{n} u_i^{\gamma_i^j}(1 - u_i)^{1-\gamma_i^j}.$$

The first step in the implementation of the EM algorithm is to compute initial estimates of the unknown parameters m, u, and p. An analyst with previous similar experience might begin by using the parameters from a similar, previous matching project as the initial parameter estimates. The algorithm is not particularly sensitive to starting values and the initial estimates of m can be guessed. Jaro [1989] states that the initial estimates of m should be greater than the corresponding initial estimates of u. Jaro [1989] used an initial estimate of .9 for each component of m in his Tampa study.

Alternatively, the analyst might use the EM algorithm itself to obtain initial estimates.

The implementation of the EM algorithm from here on involves repeated implementations of the expectation (E) step, followed in turn by the maximization (M) step until the algorithm has produced estimates that attain the desired level of precision.

For the expectation step, we replace the indicator function g_j by $\hat{g}_j$ where

$$\hat{g}_j = \frac{\hat{p}\prod_{i=1}^{n}\hat{m}_i^{\gamma_i^j}(1-\hat{m}_i)^{1-\gamma_i^j}}{\hat{p}\prod_{i=1}^{n}\hat{m}_i^{\gamma_i^j}(1-\hat{m}_i)^{1-\gamma_i^j} + (1-\hat{p})\prod_{i=1}^{n}\hat{u}_i^{\gamma_i^j}(1-\hat{u}_i)^{1-\gamma_i^j}}.$$

For the maximization step, we partition the problem into three distinct maximization problems: one for p, and one for each of the vectors m and u. Setting the partial derivatives of the complete data log-likelihood equal to zero and solving for $\hat{m}_i$ and $\hat{u}_i$, we obtain, respectively,

$$\hat{m}_i = \frac{\sum_{j=1}^{N}\hat{g}_j \cdot \gamma_i^j}{\sum_{j=1}^{N}\hat{g}_j} = \frac{\sum_{j=1}^{2^n}\hat{g}_j \cdot \gamma_i^j \cdot f\left(\gamma^j\right)}{\sum_{j=1}^{2^n}\hat{g}_j \cdot f\left(\gamma^j\right)}$$

and

$$\hat{u}_i = \frac{\sum_{j=1}^{N}\left(1-\hat{g}_j\right) \cdot \gamma_i^j}{\sum_{j=1}^{N}\left(1-\hat{g}_j\right)} = \frac{\sum_{j=1}^{2^n}\left(1-\hat{g}_j\right) \cdot \gamma_i^j \cdot f\left(\gamma^j\right)}{\sum_{j=1}^{2^n}\left(1-\hat{g}_j\right) \cdot f\left(\gamma^j\right)}$$

where $f\left(\gamma^j\right)$ is the number of times the pattern γ^j occurs in the N pairs of records. Finally, the solution for p results in the following estimate of the proportion of matched pairs:

$$\hat{p} = \frac{\sum_{j=1}^{N}\hat{g}_j}{N} = \frac{\sum_{j=1}^{2^n}\hat{g}_j \cdot f\left(\gamma^j\right)}{\sum_{j=1}^{2^n}f\left(\gamma^j\right)}.$$

Winkler [1998 – lecture notes] has additional advice for the analyst using the EM algorithm to obtain estimates of the m- and u-probabilities. The two examples that follow illustrate the methodology.

Example 9.1: Using the EM algorithm

We have two pairs of files A_1 and B_1 whose data fields include first name, surname, age, house number, and street name. We use the following initial m- and u-probabilities to get our EM algorithm started (Table 9.1).

We run the EM algorithm and, after a number of iterations, we converge to the following estimated probabilities (Table 9.2).

For this pair of files, we say that the distinguishing power of the individual fields is very good because the m-probabilities are close to one and the u-probabilities are close to zero. Moreover, the agreement weight on first name, for example, is

$$\log_2\left(\frac{P[agree\ on\ first\ name\ |\ r \in M]}{P[agree\ on\ first\ name\ |\ r \in U]}\right) = \log_2\left(\frac{.99}{.02}\right) = 5.6$$

and the disagreement weight on first name is

$$\log_2\left(\frac{1 - P[agree\ on\ first\ name\ |\ r \in M]}{1 - P[agree\ on\ first\ name\ |\ r \in U]}\right) = \log_2\left(\frac{.01}{.98}\right) = -6.6$$

Example 9.2: Another Illustration of the EM Algorithm

Here, we have two other pairs of files A_2 and B_2 whose data fields again include first name, surname, age, house number, and street name. We use the same initial m- and u-probabilities as we did in the previous example to get our EM

TABLE 9.1. Initial probabilities for files A_1 and B_1

Initial m-probabilities	Initial u-probabilities
P[agree on first name|r ∈ M] = 0.9	P[agree on first name|r ∈ U] = 0.1
P[agree on surname|r ∈ M] = 0.9	P[agree on surname|r ∈ U] = 0.1
P[agree on age|r ∈ M] = 0.9	P[agree on age|r ∈ U] = 0.1
P[agree on house number|r ∈ M] = 0.8	P[agree on house number|r ∈ U] = 0.2
P[agree on street name|r ∈ M] = 0.8	P[agree on street name|r ∈ U] = 0.2

TABLE 9.2. Final estimated probabilities for files A_1 and B_1

Final estimated m-probabilities	Final estimated u-probabilities
P[agree on first name|r ∈ M] = 0.99	P[agree on first name|r ∈ U] = 0.02
P[agree on surname|r ∈ M] = 0.92	P[agree on surname|r ∈ U] = 0.08
P[agree on age|r ∈ M] = 0.90	P[agree on age|r ∈ U] = 0.02
P[agree on house number|r ∈ M] = 0.95	P[agree on house number|r ∈ U] = 0.05
P[agree on street name|r ∈ M] = 0.95	P[agree on street name|r ∈ U] = 0.20

TABLE 9.3. Final estimated probabilities for files A_2 and B_2

Final estimated m-probabilities	Final estimated u-probabilities
P[agree on first name|r ∈ M] = 0.85	P[agree on first name|r ∈ U] = 0.03
P[agree on last name|r ∈ M] = 0.85	P[agree on last name|r ∈ U] = 0.10
P[agree on age|r ∈ M] = 0.60	P[agree on age|r ∈ U] = 0.01
P[agree on house number|r ∈ M] = 0.45	P[agree on house number|r ∈ U] = 0.01
P[agree on street name|r ∈ M] = 0.55	P[agree on street name|r ∈ U] = 0.05

algorithm started. We run the EM algorithm and, after a number of iterations, we converge to the following estimated probabilities (Table 9.3).

On the second pair of files, the distinguishing power of the individual fields is not quite as good as that of the first pair of files. In this example, the agreement weight on first name is

$$\log_2 \left(\frac{P[agree\ on\ first\ name\ |\ r \in M]}{P[agree\ on\ first\ name\ |\ r \in U]} \right) = \log_2 \left(\frac{.85}{.03} \right) = 4.8$$

and the disagreement weight on first name is

$$\log_2 \left(\frac{1 - P[agree\ on\ first\ name\ |\ r \in M]}{1 - P[agree\ on\ first\ name\ |\ r \in U]} \right) = \log_2 \left(\frac{.15}{.97} \right) = -2.7$$

Finally, we note that when the individual agreement weight is higher, it has more of a tendency to raise the total agreement weight (sum of the weights over all the fields) and gives a higher probability that a pair is a match. When a disagreement weight takes a lower negative value, it has more of a tendency to lower the total agreement weight and gives a lower probability that the pair is a match.

9.5. Advantages and Disadvantages of Using the EM Algorithm to Estimate *m*- and *u*-probabilities

In general, the EM algorithm gives us an excellent method of determining the m- and u-probabilities automatically.

9.5.1. First Advantage – Dealing with Minor Typographical Errors and Other Variations

The EM algorithm does well at obtaining probabilities that take into account the effect of minor typographical errors (e.g., Roberta versus Roburta) and other variations (e.g., first name versus nickname – Roberta versus Bobbie) in the data.

If there were no typographical error, then each matched pair would necessarily agree on its associated identifier (i.e., matching variable) with probability one. For instance, $P[agree\ first\ name \mid r \in M] = 1$ and $P[agree\ last\ name \mid r \in M] = 1$. Because there is typographical variation, it is not unusual for $P[agree\ first\ name \mid r \in M] = .9$. In other words, 10% of the truly matched pairs have a typographical variation in an identifier that will not produce an exact match. A human being can easily realize that such pairs are matches. We would like the computer to classify them as matches, too.

Another variation can occur. With one pair of files, A_1 and B_1, we might have $P[agree\ first\ name \mid r \in M] = .9$ and with another we might have $P[agree\ first\ name \mid r \in M] = .8$. We can think of the typographical variation rate (or typographical error rate) as being higher in the second file than in the first. As we saw in Examples 9.1 and 9.2, even when we start with different initial probabilities, the EM algorithm can give us "good" final estimates.

9.5.2.   Second Advantage – Getting a Good Starting Point

Another significant advantage of the EM is that it can be used as an exploratory tool for getting good initial estimates of the m- and u-probabilities in many types of files, particularly those having substantial typographical variation.

9.5.3.   Third Advantage – Distinguishing Power of m- and u- Probabilities

The EM algorithm is very good at determining the absolute distinguishing power of the m- and u-probabilities for each field and also for determining the relative distinguishing power of fields.

For instance, with individuals, names typically have better distinguishing power than addresses because there may be many individuals at the same address.

With businesses, the opposite is true. Business addresses are usually house-number-street-name types because they represent physical locations of the business entities. Typically, only one business is at a given location. Because many of the businesses have name variants – such as "John K Smith Co" on one list and "'J K S Inc" on another list – it is sometimes difficult to link business records by the name of the business. On the other hand, especially for small businesses, it is not unusual to find the address of (1) the company's accountant, (2) the place of incorporation, or (3) the owner's residence listed as the company's address. Further, a business may have multiple locations.

The EM algorithm will automatically adjust the m- and u-probabilities for both types of matching situations and will usually assign the more useful field more distinguishing power.

9.5.4. *The Main Disadvantages of Using the EM Algorithm*

The EM algorithm does not work well in situations of extreme typographical variation, however.

Some implementations of the EM algorithm may not yield good parameter estimates, with business lists, agriculture lists, and some administrative lists, for example. The main reason is the high-failure rate of name and/or address standardization/parsing software. When name or address standardization fails for a given pair that is truly a match, then we typically cannot use key identifying information (either the name or the address) to determine correctly that the pair is a true match. In this situation, the information associated with the particular match makes the pair look more like a non-match. Note that non matches typically have identifying information such as name and address that does not agree.

9.6.   General Parameter Estimation Using the EM Algorithm

Two difficulties frequently arise in applying EM algorithms of the type described in Section 9.4. First, the independence assumption is often violated (see Smith and Newcombe [1975] or Winkler [1989b]). Second, due to model misspecification, the EM algorithm or other fitting procedures may not naturally partition $A \times B$ into the two desired sets: M (the matches) and U (the non-matches). This section lays out a way to handle these difficulties.

9.6.1. *Overcoming the Shortcomings of the Schemes Described Above*

To account for dependencies between the agreements of different matching fields, an extension of an EM-type algorithm originally proposed by Haberman [1975] and later discussed by Winkler [1989a] can be applied. This extension involves the use of multi-dimensional log-linear models with interaction terms. Because many more parameters can be used within such general interaction models than with independence models, only a small fraction of the possible interaction terms may be used in the model. For example, if there are 10 fields used in the matching, the number of degrees of freedom will only be large enough to include all three-way interaction terms; with fewer matching fields, it may be necessary to restrict the model to various subsets of the three-way interaction terms. For more details on this, see Bishop, Fienberg, and Holland [1975] or Haberman [1979].

To address the natural partitioning problem, $A \times B$ is partitioned into three sets or classes: C_1, C_2, and C_3. The three-class EM algorithm works best when we are matching persons within households and there are multiple persons per household.

When appropriate, two of the three classes may constitute a partition of either M or U; for example, we might have $M = C_1$ and $U = C_2 \cup C_3$. When both name and address information are used in the matching, the two-class EM algorithm tends to partition $A \times B$ into (1) those agreeing on address information and (2) those disagreeing. If address information associated with many pairs of records is indeterminate (e.g., Rural Route 1 or Highway 65 West), the three-class EM algorithm may yield a good partition because it tends to partition $A \times B$ into (1) matches at the same address, (2) non-matches at the same address, and (3) non-matches at different addresses.

The general EM algorithm is far slower than the independent EM algorithm because the M-step is no longer in closed form. The use of a variant of the Expectation-Conditional Maximization (ECM) algorithm may speed up convergence (see Meng and Rubin [1993] or Winkler [1992]). The difficulty with general EM procedures is that different starting points often result in different limiting solutions. A recommended approach is to use the independent EM algorithm to derive a starting point for the general scheme as this usually produces a unique solution.

9.6.2. Applications

When matching individuals across two household surveys, we typically have (1) fields on individuals such as first name and age and (2) fields on the entire household such as surname, house number, street name, and phone number. A general two-class EM algorithm will partition a set of record pairs according to the matching variables. Since there are more household fields than fields on individuals, the household fields usually overwhelm person variables in the two-class EM scheme and partition the set of pairs into those at the same address (within the same household) and those not at the same address. In this case, better results are usually obtained by using the three-class EM algorithm. As noted above, the three-class EM algorithm tends to partition the record pairs into three classes: matches within households, non-matches within households, and non-matches outside of households. The three-class EM algorithm works by estimating the probabilities associated with each of these three classes and then estimating the probabilities associated with the non-matches by combining the two non-matching classes.

Example 9.3: Using files with good distinguishing information

This example, which comes from Winkler [2000], entails matching two files of census data from Los Angeles. Each file contained approximately 20,000 records, with the smaller file having 19,682 records. The observed counts for $1024 (= 2^{10})$ agree/disagree patterns on 10 fields were used to obtain most parameter estimates. These fields were first name, middle initial, house number, street name, unit (or apartment) number, age, gender, relationship to head of household, marital status, and race. Frequency-based weights were created for both surname and first name. The basic weight for last name was created via an ad hoc procedure that attempted to account for (1) typographical error and (2) the number of pairs in the subset of pairs on which the matching was performed. A total of 249,000 pairs of records that agreed on the first character of the last name and the Census block number were considered. As there can be at most 20,000 matches here, it is not computationally practical to consider counts based on all of the 400 million pairs in the product space. Based on prior experience, Winkler was confident that more than 70% of the matches would be in the set of 249,000 pairs. The results are summarized in Table 9.4.

Winkler gave several reasons why frequency-based matching performed better here than basic matching that uses only agree/disagree weights. First, the frequency-based approach designated 808 pairs of records as matches that the basic matcher merely designated as possible matches. These pairs of records were primarily those having (1) rare surnames, (2) rare first names, and (3) a moderate number of disagreements on other fields. Second, the frequency-based approach designated 386 pairs of records as possible matches that the basic matcher designated as non-matches. These pairs of records were primarily those having (1) rare last names and (2) few, if any, agreements on other fields.

Example 9.4 – Using files with poor distinguishing information

This example, also from Winkler [2000], entails matching two files of census data from St. Louis. The larger file contains 13,719 records while the smaller one has 2,777 records. The smaller file was obtained by merging a number of administrative data files. The observed counts for $128 \left(= 2^7\right)$ agree/disagree patterns on seven fields were used to obtain most parameter estimates. These fields were first name, middle initial, address, age, gender, telephone number, and race. Frequency-based weights were created for both last name and first name. The basic weight for last name was created via an ad hoc procedure that attempted to account for (1) typographical error and (2) the number of pairs in the subset of pairs on which the matching was performed. A total of 43,377 pairs of records that agreed on the Soundex code (see Chapter 10) of the last name were considered. The results are summarized in Table 9.5.

Neither matching scheme performed well in this example. Each scheme classified approximately 330 records as matches. The file used to obtain additional information about black males between the ages of 18 and 44 had many missing data fields. The middle initial, telephone number, and race fields were missing on 1,201, 2,153 and 1,091 records, respectively. The age and

TABLE 9.4. Comparing basic estimates to frequency-based estimates (data from Los Angeles)

Basic approach	Frequency-based approach			Total
	Match	Possible match	Non-match	
Match	12,320	128	7	12,455
Possible match	808	2,146	58	3,012
Non-match	8	386	3,821	4,215
Total	13,136	2,660	3,886	19,682

TABLE 9.5. Comparing basic estimates to frequency-based estimates (data from St. Louis)

Basic approach	Frequency-based approach			Total
	Match	Possible match	Non-match	
Match	305	21	2	328
Possible match	15	142	0	157
Non-match	2	106	2,184	2,292
Total	332	269	2,186	2,777

address fields were also frequently incorrect. The frequency-based matching scheme designated 269 record pairs as possible matches versus 157 for the basic matching scheme. These 269 record pairs typically had (1) a relatively rare first name, (2) a relatively rare last name, and (3) few, if any, agreements on other fields.

9.7.   Where Are We Now?

In this chapter, we have discussed a variety of schemes for estimating the parameters of the Fellegi–Sunter record linkage model. We also discussed the advantages and disadvantages of these schemes. We concluded the chapter with a pair of examples in which we compared frequency-based record linkage to basic record linkage. In the next four chapters, we discuss a number of techniques that can be used to enhance record linkages.

10
Standardization and Parsing

The purpose of record linkage, as noted in previous chapters, is to find (1) pairs of records across files that correspond to the same entities (e.g., individuals or businesses) or (2) duplicates within a single file. If unique identifying numbers are not available, then the name and/or address are usually used. These names and addresses can be considered to be strings of characters or *character strings*.

The first step in comparing such character strings is to identify commonly used terms such as road, street, mister, or junior. These are then converted, if necessary, to a standard form or abbreviation. We use *standardization* to refer to this conversion of common terms to a standard form or abbreviation. Common terms such as Doctor, Missus, etc. (Dr, Mrs.) or Street, Drive, etc., or Corporation, Company etc. (Corp., Co.) can be relatively easy to find and standardize.

The second step is to partition the character string, into its component parts and to place these parts into identifiable fields. In the terminology of record linkage, we use the term *parse* to indicate this type of partition. When parsing these strings it is often useful to make use of the results of the standardization process. Specifically, the standardized words can serve as keywords into special routines for parsing. For instance, if you find the term "Company" in the string, then you are likely dealing with the name of a business and so want to call on a computer subroutine that parses business addresses. Similarly, Post Office Box is different from Rural Route Box Number; and Rural Route Box Number is different from House Number Street Name.

Standardization and parsing are used to facilitate matching and need to be completed before comparing the records for record linkage purposes.

Standardization is discussed in Section 10.2, parsing in Section 10.3. Unfortunately, it is not always possible to compare/match two strings exactly (character-by-character) because of typographical error and/or variations in spelling. In record linkage, we need a function that represents approximate agreement, with complete agreement being represented by 1 and partial agreement being represented by a number between 0 and 1. We also need to incorporate these partial agreement values into the weighting process – that is, to adjust the weights (likelihood ratios of the Fellegi–Sunter scheme). Such functions are called *string comparator metrics* and are discussed in Chapter 13. Additionally, we sometimes encode street names, surnames, and/or given names to facilitate comparison. Two

TABLE 10.1. Which pairs represent the same person?

Name	Address	Age
John A Smith	16 Main Street	16
J H Smith	16 Main St	17
Javier Martinez	49 E Applecross Road	33
Havier Matreenez	49 Aplecross Raod	36
Bobbie Sabin	645 Reading Aev	22
Roberta Sabin	123 Norcross Blvd	
Gillian Jones	645 Reading Aev	24
Jilliam Brown	123 Norcross Bvd	43

widely used coding schemes, SOUNDEX and NYSIIS, used for such purposes, are described in Chapter 11. In Chapter 12 we describe a scheme known as blocking that reduces the number of record pairs we need to compare.

We begin this chapter by looking at some examples of the problems for which standardization and parsing are likely to be helpful.

Example 10.1: Which pairs are true?

We consider the following pairs of entries taken from a variety of lists (Table 10.1).

The question that arises is, Which of the pairs above represent the same person? The answer is that we need to see each of the lists in its entirety. We need to know the context in which the entries appear. For instance, with the Sabin example, the first entry might be from a list of individuals in a college glee club and the second might be from a list of known graduates of the college. Concerning the Jones/Brown example, the first entry might be from a list of medical students at a particular university 20 years ago and the second list might be a current list of practicing physicians who had graduated from the university's medical school. Relatively inexperienced clerical staff should be able to figure out the first two examples easily. We would like to have computer software that does some of the things a person can do and also does the things that a computer does well.

Example 10.2: Which pairs are real, again?

We consider the names on two records as shown in Table 10.2

TABLE 10.2. Names as originally entered on data record

Record	Name
I	Smith, Mr. Bob, Jr.
II	Robert Smith

TABLE 10.3. Standardized names

Record	Standardized name
I	Smith, Mr. Robert, Jr.
II	Robert Smith

Each of the names can be considered to be a single string. The standardized versions of these names are shown in Table 10.3.

Again, each of the standardized names is simply a string. If we then parse the standardized names, we might end up with the following:

TABLE 10.4. Parsed and standardized names

Record	Prefix	First name (standardized)	Surname	Suffix
I	Mr.	Robert	Smith	Jr.
II		Robert	Smith	

We analyze the situation as follows. We began with the names "Smith, Mr. Bob Jr." and "Robert Smith" appearing on two records (call them I and II, respectively). "Smith, Mr. Bob Jr." is composed of a last name "Smith," a title "Mr.", an abbreviation of a first name "Bob," and a suffix, "Jr.". "Robert Smith" is composed of a (unabbreviated) first name, "Robert", and a last name "Smith." Without the ability to identify the different parts of a name and to place standard spellings or abbreviations for these parts into fixed fields, one could only compare the two character strings on a letter-by-letter basis to see if they were identical. Comparing "S" to "R," "m" to "o," "i" to "b," etc., would cause the computer to believe that the two name fields were different. With the ability to parse and standardize the names, the two name fields would appear as shown in Table 10.4. The computer could then compare each corresponding part. It would discover that it had two missing values and two perfect agreements, rather than a disagreement on a single long string.

10.1.   Obtaining and Understanding Computer Files

What files can we use for research studies? What work needs to be done so that the information on a file can be used for matching? What work needs to be done so that we can match records across files? We need an annotated layout. We need to know which records are "in-scope," as some files contain copies of records with a previous address and a status code identifying such records as duplicates. We need to determine the proportion of records that is blank.

Observation: Prior to matching, files must be put in common forms (made homogenous) so that corresponding fields can be compared. It can take more time (moderately skilled human intervention) to pre-process a small local list than it does to pre-process a very large, well-documented, well-maintained national list.

Rule of Thumb: Get the fewest lists possible to use in updating or creating a merged file. Record linkage errors are often cumulative. If a record is erroneously contained in a file (because it is out of scope or a duplicate), then it may be added to another file (during updating) and increase error (duplication) in the updated file.

10.2. Standardization of Terms

Before a character string (such as a name or an address) is parsed into its components parts and these parts are placed into identifiable fields, each of these fields may need to be *standardized* or converted into a standard form or abbreviation. This *standardization* process (also known as data cleansing or attribute-level reconciliation) is used before performing record linkage in order to increase the probability of finding matches. Without standardization, many true matches would be erroneously designated as non-matches because the common identifying attributes would not have sufficient similarity.

The basic ideas of standardization are concerned with standardization of spelling, consistency of coding, and elimination of entries that are outside of the scope of the area of interest.

10.2.1. Standardization of Spelling

Replace spelling variations of commonly occurring words with a common consistent spelling. For example, replace "Doctor" or "Dr" by "Dr"; replace nicknames such as "Bob" and "Bill" by "Robert" and "William," respectively; replace "rd" or "road" by "rd"; and replace "company," "cmpny," or "co" by "co." We note that the last example is dependent on the context of the term as "co" might refer to county or even Colorado.

10.2.2. Consistency of Coding

Standardize the representation of various attributes to the same system of units or to the same coding scheme. For example, use 0/1 instead of M/F for a "gender" attribute; replace "January 11, 1999" or "11 January 1999" by "01111999" (i.e., MMDDYYYY) or "19990111" (i.e., YYYYMMDD).

10.2.3. Elimination of Entries That Are Out of Scope

If a list of registered voters contains entries for individuals that are deceased or relocated, these need to be edited in an appropriate fashion – that is, by adding status flags or codes, as required.

An electric utility company list of customers is a good source of residential addresses. However, such a list may contain some commercial addresses as well. So, if the goal is to compile a list of residential addresses, it is necessary to eliminate the out-of-scope commercial addresses from the source list.

Next, we consider a case involving two lists. The first is a list of the residents of a small, contiguous portion of a large urban area, while the second list is that of the entire urban area. We wish to extract telephone numbers from the larger list to augment the information on the smaller list. Before linking records, we want to eliminate the records on the larger list that are not in the geographic area represented by the smaller list. This might be accomplished by using the Zip Codes on the record entries.

10.2.4. *Perform integrity checks on attribute values or combinations of attribute values*

In this regard, we can use the types of schemes described earlier in Chapter 5.

10.2.5. *Final Thoughts on Standardization*

Standardization methods need to be specific to the population under study and the data collection processes. For example, as noted earlier, the most common name misspellings differ based upon the origin of the name. Therefore, standardization for Italian names optimized to handle Italian names and Latin origin will perform better than generic standardization.

10.3. Parsing of Fields

It is not easy to compare free-form names and addresses except possibly manually. Parsing partitions a free-form string (usually a name or an address) into a common set of components that can be more easily compared by a computer. Appropriate parsing of name and address components is the most critical part of computerized record linkage. Parsing requires the identification of the starting and ending positions of the individual components of the string. For names, we usually need to identify the locations of the first name, middle initial, and last name. For addresses, we frequently need to identify the locations of the house number and the street name. Without effectively parsing such strings, we would erroneously designate many true matches as non-matches because we could not compare common identifying information. For specific types of establishment lists, the drastic effect of parsing failure has been quantified (see Winkler [1985b and 1986]). DeGuire [1988] presents an overview of ideas needed for parsing (as well as standardizing) addresses. Parsing of names requires similar procedures.

TABLE 10.5. Examples of name parsing

Standardized name	Parsed							
	Pre	First	Middle	Last	Post1	Post2	Bus1	Bus2
Dr. John J Smith MD	DR	John	J	Smith	MD			
Smith DRY FRM				Smith			DRY	FRM
Smith & Son ENTP				Smith		Son	ENTP	

10.3.1. Parsing Names of Individuals

In the examples of Table 10.5, the word "Smith" is the name component with the most identifying information. "PRE" refers to a prefix, "POST1" and "POST2" refer to postfixes, while "BUS1" and "BUS2" refer to commonly occurring words associated with businesses. While exact, character-by-character comparison of the standardized but unparsed names would yield no matches, use of the sub-component last name "Smith" might help to designate some pairs as matches. Parsing algorithms are available that can deal with either last-name-first types of names such as "Smith, John" or last-name-last types such as "John Smith." None are available that can accurately parse both types of names within a single file.

More generally, in order to make the matching of records on individuals efficient, we need high-quality, time-independent identifiers on these individuals. These identifiers include given name, middle initial, last name, maiden name (if appropriate), Social Security number, date of birth (preferably in the format of MMDDYYYY), and city of birth.

10.3.2. Parsing of Addresses

Humans can easily compare many types of addresses because they can associate corresponding components in free-form addresses. To be most effective, matching software requires corresponding address subcomponents in specified locations. As the examples in Table 10.6 show, parsing software partitions a free-form address into a set of components each of which is in a specified location.

TABLE 10.6. Examples of address parsing

Standardized address	Parsed									
	Pre2	HSNM	STNM	RR	Box	Post1	Post2	Unit1	Unit2	Bldg
16 W Main ST APT 16	W	16	Main			ST		16		
RR 2 BX 215				2	215					
Fuller BLDG SUITE 405									405	Fuller
14588 HWY 16 W		14588	Hwy 16				W			

TABLE 10.7. Pairs of names referring to the same business entity

Name	Explanation
John J Smith ABC Fuel Oil	One list has the name of the owner while the other list has the name of the business.
John J Smith, Inc J J Smith Enterprises	These are alternative names of the same business.
Four Star Fuel, Exxon Distributor Four Star Fuel	One list has both the name of the independent fuel oil dealer and the associated major oil company.
Peter Knox Dairy Farm Peter J Knox	One list has the name of business while the other has the name of the owner.

TABLE 10.8. Names referring to different businesses

Name	Explanation
John J Smith Smith Fuel	Similar names but different companies
ABC Fuel ABC Plumbing	Identical initials but different companies
North Star Fuel, Exxon Distributor Exxon	Independent affiliate and company with which affiliated

10.3.3. Parsing Business Names

The main difficulty with business names is that even when they are parsed correctly, the identifying information may be indeterminate. In each example of Table 10.7, the pairs refer to the same business entities that might be in a list frame constructed for a sample survey of businesses.

In Table 10.8 each pair refers to different business entities that have similar components.

Because the name information in Tables 10.7 and 10.8 may not be sufficiently accurate to determine match status, address information or other identifying characteristics may have to be obtained via clerical review. If the additional address information is indeterminate, then at least one of the establishments in each pair may have to be contacted.

What information do we need to match individual business enterprises efficiently? We need information such as the business's name, the Zip Code of the business's headquarters, the Standard Industrial Classification (SIC) Code, or

North American Industry Classification System (NAICS) code of the business, as well as additional quantitative information.

10.3.4.   Ambiguous Addresses

A postal address should not merely be regarded from a syntactic point of view. Its semantic content (i.e., its meaning) must be examined as well. Sometimes, a recorded address could potentially represent two or more physical locations. In order to resolve this potential ambiguity more knowledge may be required to exclude non-existent locations and to determine the correct location. Unfortunately, this does not always lead to a resolution and so we might still end up with an ambiguity that we cannot resolve. Consider the following address:

61 1 4-th Street, N.W., Washington, D.C.

Does this represent "611 4-th Street, N.W." or "61 14-th Street, N.W."? What about the address

976 Fort St John BC?

Which of the following does the last address represent:

Apt 976, Fort St-John, BC,
Apt 976 Fort, St-John, BC, or
Apt 976 Fort ST, John, BC.?

10.3.5.   Concluding Thought on Parsing

Finally, no matter how good the software becomes, the unsolvable and the non-existent addresses will remain a problem and should be followed up manually.

10.4.   Where Are We Now?

In this chapter we discussed two techniques – standardization and parsing – that can be used to enhance record linkages. In Chapters 11–13, we discuss other techniques that also enhance record linkages.

11
Phonetic Coding Systems for Names

Phonetic coding systems use the way words or syllables are pronounced when spoken to help reduce minor typographical errors. Soundex and the New York State Identification and Intelligence System (NYSIIS) are two widely used phonetic schemes for encoding names. NYSIIS results in substantially more codes than does Soundex, and is harder to describe. Although NYSIIS provides many more codes, individual NYSIIS and Soundex codes associated with commonly occurring surnames such as Smith, Johnson, Brown, and Martin have approximately the same number of records associated with them. Neither Soundex nor NYSIIS can deal with most insertions, deletions, or transpositions of consonants. The primary value of both Soundex and NYSIIS in record linkage is to assist in bringing together records – that is, in *blocking* records. In fact, Jaro [1989] suggests using the Soundex version of names of individuals only as a blocking variable. He argues that using it for other purposes can create problems because many different names have the same Soundex code. In Chapter 12, we provide a detailed description of blocking and include several examples.[1]

11.1.   Soundex System of Names

Soundex is an algorithm devised to code people's last names phonetically by reducing them to the first letter and up to three digits, where each digit represents one of six consonant sounds. This facilitates the matching of words (e.g., names of individuals or names of streets) by eliminating variations in spelling or typographical errors. Such schemes might be usefully employed with airline reservations systems or other applications involving people's names when there is a good chance that the name will be misspelled due to poor handwriting or voice transmission. For example, Soundex would treat "Smith," "Smithe," and "Smyth" as the same name. The standard Soundex algorithm works best on European last names. Variants have been devised for names from other continents/cultures/ethnic backgrounds.

---

[1] The string comparator metrics discussed in Chapter 13 are much better than coding schemes at comparing two strings that are brought together. However, string comparator metrics are of little use in blocking records together (except with relatively small files).

11.1.1. History of Soundex

Beginning in 1935, the Social Security Administration offered old-age pensions to Americans who could show they were at least 65 years of age. Unfortunately, many of the seniors eligible for such pensions had neither birth certificates nor any other means of establishing their date of birth. For example, a senior reaching her 65th birthday during 1935 would have been born in 1870. Because there were few statewide or even countywide birth registration systems in operation in 1870, the Social Security Administration asked the US Congress to help provide an alternative method of establishing a person's age from official records.

The solution was to obtain the requisite birth dates from US Decennial Censuses. The job was assigned to the Works Progress Administration (WPA). However, the Federal government's first need was for an index to enable it to get easy access to the records of the Decennial Censuses. The system selected to index names from the censuses was called "Soundex" – a coding system that codes names according to their sound.

11.1.2. History of Census Bureau Soundex Indices

According to Dollarhide,[2] The Soundex indexing system "was used to create heads of household name indexes to the 1880, 1900, 1910, 1920, and 1930 [Decennial] Censuses."

"The 1880 Soundex index was compiled for all states but lists only families in which a child of 10 years or younger was included. The 1900 and 1920 indexes are complete indexes to all heads of household for all states. The 1930 Soundex was compiled only for twelve southern states only."

"In the 1910 census index, only 21 states were indexed. Of these, 15 of the states were indexed under the name 'Miracode,' and 5 states were done using the Soundex name. The coding of last names in Soundex and Miracode was identical. In fact, the only difference between Soundex and Miracode was in the citation of a family's position on a census page."

"All of the Soundex indexes were originally hand-written on 3″ × 5″ index cards, each card showing a head of household by full name and a list of all other persons residing in a household. Persons with a different surname than the head of household usually had a separate index card, coded under their own surname, as well as the one for the household in which they were listed. Included was each person's name, age, relationship to the head of household, nativity [i.e., place of birth], and a reference to the location in the census schedules where that family appears."

---

[2] The material cited in this section was taken from some Internet sites that no longer exist. The author of that material is Dollarhide who is also the author of *The Census Book* that is available on the Internet at www.heritagequestonline.com/prod/genealogy/html/help/census_book.html

"The Works Progress Administration (WPA) compiled the 1880, 1900, 1920, and 1930 census indexes in the late 1930s, while the Age Search group[3] of the Census Bureau compiled the 1910 Soundex/Miracode index in 1962. Only the 1910 Miracode names were entered into computers for sorting purposes. This was done by keypunch in 1962, and the data [were processed by mainframe computers, with the output consisting of a printed strip] for each entry. The microfilmed images of the 1880, 1900, 1920 and 1930 Soundex are handwritten cards, while the images of the 1910 Miracode are from computer-generated printed strips. (For the five 1910 Soundex states, the originals were the same type of 3″ × 5″ index card as the other Soundex indexes; and the bottom of each printed card has the year 1962 on it.)"

Every Soundex index is now available to the public.

11.1.3. Soundex Indexes

"In all Soundex indices, the original cards were sorted by the Soundex code for the head of household's surname, then alphabetized by each person's first name. As a result, all [surnames] with the same Soundex codes are interfiled." For example, all last names with the code L000 (e.g., Lee, Leigh, Low, Law, Liem, and Lieh) are interfiled. So if you know the Soundex code and the first name of an individual, you can go directly to that individual's index card.

11.1.4. Soundex Coding Rules

The algorithm below implements a modern version of the Soundex coding system, a technique that Knuth [1998] attributes to Margaret K. Odell and Robert C. Russell [see *US Patents 1261167* (1918) and *1435663* (1922)].

SOUNDEX ALGORITHM

Step1: The first letter of a last name retains its alphabetic designation.
Step 2: In other than the first position, the letters a, e, i, o, u, y, w, and h are not coded at all. (This eliminates the vowels and the mostly silent sounding letters from the coding scheme.)
Step 3: The remaining consonants – the hard consonants – are coded according to the table below.

---

[3] During that period the Census Bureau set up a special "Age Search" group tasked with using "the Soundex Indexes to locate a person in one of the censuses. The Age Search group researched applications for Social Security pensions from seniors who could not prove their age. Upon finding someone in the census, the group would record the person's name, [place of birth], and age information and then issue a substitute for a birth certificate." "The Age Search group still exists, and its primary function remains" to enable individuals to verify their age via information obtained from prior censuses.

Coding Guide

Code	Key letters and equivalents
1	b, p, f, v
2	c, s, k, g, j, q, x, z
3	d, t
4	l
5	m, n
6	r

Step 4: If two or more letters with the same code were adjacent in the original name (prior to Step 1), or adjacent except for intervening h's and w's, then only the first letter is coded.

Step 5: Every Soundex code must consist of one alphabetic character followed by three digits. So, if there are less than three digits, an appropriate number of zeroes must be added; if there are more than three digits, then the appropriate number of the rightmost digits must be dropped.

Example 11.1:

To illustrate Steps 1, 2, and 5, we consider names such as Lee and Shaw. A name consisting of all vowels after the first letter, such as Lee, would be coded as L000 because the first letter, "L", is always retained, the two vowels, both "e", are dropped, and three zeroes are appended to attain the required three digits. Similarly, Shaw would be coded as S000 because the first letter, "S", is retained, the vowel, "a", is dropped as are the "mostly silent sounding letters" "h" and "w", and three zeroes are again appended to attain the required three digits.

Example 11.2:

To illustrate Step 4, we consider names such as Gauss, Cherry, and Checker. Gauss is coded as G200 because the two vowels are dropped, the double "s" is treated as a single letter, and two zeroes are appended to attain the requisite three digits. Similarly, Cherry is coded as C600 because the two vowels and the "h" are dropped, the double "r" is treated as a single letter, and two zeroes are appended to attain the requisite three digits. Finally, Checker is coded as C260 because the two "e"s and the "h" are dropped, the pair of letters "ck" is treated as a single letter since they have the same code (namely, "2"), the "r" is coded as a "6," and one zero is appended to attain the requisite three digits.

Example 11.3:

For a more complex example, we consider a name such as "Coussacsk." Here, the three vowels are dropped, the double-letter sequence "ss" is coded as "2" and the three-letter sequence "csk" is also coded as "2" because these three consecutive letters all have the same code of "2." So, "Coussacsk" is coded as C220.

11.1.5. Anomalies with Soundex

Some names that are closely related are coded differently. For example, while "Lee" is coded as L000, "Leigh" is coded as L200; "Rogers" is coded as R262 while "Rodgers" is coded as R326; and "Tchebysheff" is coded as T212 while "Chebyshev" is coded as C121.

On the other hand, some unrelated names have identical codes in Soundex. For example, both "Lee" and "Liu" are coded as L000; both "Gauss" and "Ghosh" are coded as G200; and both "Wachs" and "Waugh" are coded as W200.

Another issue is how to treat names such as Lloyd, van Buren, or von Munching.

Despite the problems noted above, Knuth [1998] concludes that "by and large the Soundex code greatly increases the chance of finding a name in one of its disguises."

11.2. New York State Identification and Intelligence System (NYSIIS) Phonetic Decoder

In this section, we present a seven-step procedure for implementing the NYSIIS coding scheme as Taft [1970] originally proposed it. In the first step, the initial letter(s) of a surname are examined and altered as necessary. In the second step, the same is done for the last letter(s) of the surname. In Step 3, the first letter of the NYSIIS coded surname is established. Steps 5 and 6 constitute an iterative procedure for creating the remaining letters of the NYSIIS-coded surname. In this iterative scheme, we begin with the second letter of the altered surname and scan each letter of the remaining letters of the surname using an imaginary "pointer." In Step 5, one or more of the letters of the coded surname are established via a set of "rules." The rules are reapplied each time the "pointer" is moved to the next letter of the name. In Step 7, the end portion of the NYSIIS code just created is subjected to a further check and changed as necessary.

In an in-depth comparative study of five coding systems, Lynch and Arends [1977] judged a "modified" NYSIIS coding scheme to be the best surname coding system because:

It (1) placed variations of a given surname in the same code, (2) limited the size of each code, and (3) created codes that contain few dissimilar names.

The individual steps of the NYSIIS coding scheme are as follows, with coding beginning in Step 3:

Step 1: Change the initial letter(s) of the surname as indicated in the table below:

Changing the initial letter(s) of the surname	
Original letter(s)	Altered letter(s)
MAC	MCC
KN	NN
K	C
PH	FF
PF	FF
SCH	SSS

Step 2: Change the last letter(s) of the surname as indicated in the table below.

Original letter(s)	Altered letter
EE	Y
IE	Y
DT	D
RT	D
RD	D
NT	D
ND	D

Step 3: The first character of the NYSIIS-coded surname is the first letter of the (possibly altered) surname.

Step 4: Position the "pointer" at the second letter of the (possibly altered) surname.

Step 5: (Change the current letter(s) of the surname – i.e., the one at the present position of the "pointer".) Execute exactly one of the following operations, proceeding from top to bottom:

(a) If blank, go to Step 7.
(b) If the current letter is "E" and the next letter is "V," then change "EV" to "AF."
(c) Change a vowel ("AEIOU") to "A."
(d) Change "Q" to "G."
(e) Change "Z" to "S."
(f) Change "M" to "N."
(g) If the current letter is the letter "K," then change "K" to "C" unless the next letter is "N." If "K" is followed by "N," then replace "KN" by "N."
(h) Change "SCH" to "SSS."
(i) Change "PH" to "FF."
(j) If "H" is preceded by or followed by a letter that is *not* a vowel (AEIOU), then replace the current letter in the surname by the preceding letter.

(k) If "W" is preceded by a vowel, then replace the current letter in the surname with the preceding letter.

Step 6: The next character of the NYSIIS code is the current position letter in the surname after completing Step 5 (but omitting a letter that is equal to the last character already placed in the code).

   After putting a character into the code, move the pointer forward to the next letter of the surname. Then return to Step 5.

Step 7: (Change the last character(s) of the NYSIIS-coded surname.) If the last two characters of the NYSIIS-coded surname are "AY," then replace "AY" by "Y." If the last character of the NYSIIS-coded surname is either "S" or "A," then delete it.

Example 11.4:

We illustrate the use of the NYSIIS coding system in the examples summarized in the following table:

Examples of use of NYSIIS coding system[4]

Surname	NYSIIS-coded surname
Brian, Brown, Brun	Bran
Capp, Cope, Copp. Kipp	Cap
Dane, Dean, Dent, Dionne	Dan
Smith, Schmit, Schmidt	Snat
Trueman, Truman	Tranan

11.3. Where Are We Now?

In this chapter we discussed two coding schemes that can be used to enhance record linkages. These schemes have primary application in blocking – the subject of our next chapter.

---

[4] All of these examples are taken from Newcombe [1988].

12
Blocking

Suppose that for two files, *A* and *B*, of average size, the number of records in the product space $A \times B$ is too big for us to consider all possible record pairs. Because (1) only a small portion of the pairs in $A \times B$ are matching records and (2) there are 2^n possible comparison configurations involving *n* fields, drawing record pairs at random would require a sample size approaching all record pairs (for typical applications) to obtain sufficient information about the relatively rare matching pairs.

Blocking is a scheme that reduces the number of pairs of records that needs to be examined. In blocking, the two files are partitioned into mutually exclusive and exhaustive blocks designed to increase the proportion of matches observed while decreasing the number of pairs to compare. Comparisons are restricted to record pairs within each block. Consequently, blocking is generally implemented by partitioning the two files based on the values of one or more fields. For example, if both files were sorted by Zip Code, the pairs to be compared would only be drawn from those records whose Zip Codes agree. Record pairs disagreeing on Zip Code would not be on the same sub-file and hence would be automatically classified as non-matches (i.e., pairs in U).[1]

To illustrate this concept further, we consider an age variable. If there are 100 possible ages, then this variable would partition the database into 100 subsets. The first subset would be all of the infants less than one year of age. The second subset would be all those between ages one and two, etc. Such subsets are known as *blocks* or *pockets*.

Example 12.1: Calculating the number of non-matches

Suppose file A has 2,000 records and file B has 3,000 records. Further, suppose that there are no duplicate records within either file.

(1) How many possible record pairs, (a, b), are there in which $a \in A$ and $b \in B$?
(2) What is the maximum number of matches (a, b) where $a \in A$ and $b \in B$?

---

[1] In census studies, this grouping variable frequently identifies a small geographic area. This may well be the reason that this scheme is known as "blocking."

(3) If the number of matches attains its maximum value as computed in (2), what is the corresponding number of non-matches?

Solution

(1) The number of pairs in $A \times B$ is $2,000 \times 3,000 = 6,000,000$ – a large number indeed!
(2) The maximum number of matches is 2,000, the number of records in the smaller file.
(3) The number of non-matches is $6,000,000 - 2,000 = 5,998,000$. Thus, even the maximum number of matches is a relatively small proportion of the total number of pairs within $A \times B$.

Example 12.2: Calculating the number of pairs within an age group

Suppose that file A of Example 12.1 had 20 people at each of the ages from 0 to 99 and that file B of Example 12.1 had 30 people at each of the ages from 0 to 99. Suppose that a block consisted of all those people in either file A or File B at age 35. How many possible record pairs, $(a, b) \in A \times B$, are there in the block in which the recorded age on both records a and b is "35"?

Solution

There are 20 such individuals from file A and 30 from file B, so the answer is $20 \times 30 = 600$.

Hence, to do all pairwise comparisons within each of the one hundred blocks would require only $100 \times 600 = 60,000$ comparisons – only 1% of the 6,000,000 comparisons required to compare all of the records of file A to all of the records of file B without blocking.

12.1.   Independence of Blocking Strategies

Blocking causes all records having the same values on the blocking variables to be compared. One consequence of this is that records which disagree on the blocking fields will be classified as non-matches. In particular, if the age field is used as one of the blocking variables and a record of file A has an erroneous age recorded, then that record would not be matched to the appropriate record of file B (unless the record on file B had a corresponding error). To circumvent this situation, we use multiple passes of the data.

Suppose that we ran a match in which the Soundex code of the last name was the sole blocking field on the first pass and the postal code of one's primary residence was the sole blocking variable on the second pass. If a record did not match on the first pass because the age was incorrect, then that record might still match on the second pass. Further passes could be made until the analyst felt that it was unlikely that matches would be missed because of errors in the blocking fields. Errors on three specified blocking fields are unlikely.

The blocking strategies for each pass should be independent to the maximum extent possible. For example, if a pair of files had last name, first name, gender, and birthdate (year, month, and day) fields, then the first pass could be blocked on last name, gender, and year of birth. The second pass could be blocked on first name and day and month of birth. Consequently, matching records having errors in last name (e.g., a maiden and a married surname), for example, would not be matched in the first pass, but would probably be matched in the second pass.

12.2.   Blocking Variables

To identify the matches efficiently, blocking variables should (1) contain a large number of possible values that are fairly uniformly distributed and (2) have a low probability of reporting error. Smaller blocks are more effective than larger ones. It is preferable to use highly restrictive blocking schemes in the first pass. The records that match in the first pass can then be removed from the databases. Because most records that match will match on the first pass, less restrictive schemes can then be used in subsequent passes to identify additional matches.

For example, gender, by itself, is a poor blocking field because it only partitions the file into two sub-files. Even a field such as age is not a particularly good blocking field because (1) it is usually not uniformly distributed, (2) it is usually more effective to partition large files into more than 100 subsets, and (3) it is time dependent so it often disagrees on pairs of records that should match. You want to use the blocking to partition the database into a large number of small segments so that the number of pairs being compared is of a reasonable size.

Also, fields subject to a high probability of error should not be used as blocking fields. For example, apartment number is widely misreported or omitted and hence would not make a good blocking field. Accuracy is crucial because a failure of two matching records to agree on the values of the blocking field would cause the records to be placed in different blocks and thus have no chance to be matched.

Blocking is a trade-off between computation cost (examining too many record pairs) and false non-match rates (classifying matching record pairs as non-matches because the records are not members of the same block). Multiple-pass matching techniques, using independent blocking fields for each run, can minimize the effect of errors in a set of blocking fields.

12.3.   Using Blocking Strategies to Identify Duplicate List Entries

Winkler [1984] describes a study that evaluated methodologies for accurately matching pairs of records – within a single list of records – that are not easily matched using elementary comparisons.

The empirical database he used was a part of a list frame that the Energy Information Administration (EIA) of the US Department of Energy employed to conduct sample surveys of sellers of petroleum products. The list was constructed from (1) 11 lists maintained by the EIA and (2) 47 State and industry lists containing roughly 176,000 records. Easily identified duplicates having essentially similar name and address fields were deleted, reducing the merged file to approximately 66,000 records. This was more suitable for Winkler's study because he was concerned with accurately identifying 3,050 other duplicate records that only had somewhat similar names and addresses, and so were more difficult to identify.

Winkler [1984] used the following five blocking criteria in his study:

1. The first three digits of the Zip Code field and the first four characters of the NAME field in its original form – that is, not parsed.
2. The first five digits of the Zip Code field and the first six characters of the STREET name field.
3. The 10-digit TELEPHONE number.
4. The first three digits of the Zip Code field and the first four characters of the longest substring in the NAME field.
5. The first 10 characters of the NAME field.

Table 12.1 summarizes the results of applying these five criteria to the list of interest. These criteria were the best of several hundred considered.

As we can see from Table 12.1, the best blocking criterion (criterion 4) failed to identify 702 or 23.0% of the 3,050 duplicates on the list. The reason criterion 4 works best is that the NAME field does not have subfields (generally words)

TABLE 12.1. Number of matches, erroneous matches, and erroneous non-matches using a single blocking criterion

Blocking criterion	Matched with correct parent (match)	Matched with wrong parent (erroneous match)	Not matched (erroneous non-match)	Percentage of erroneous non-matches
1	1,460	727	1,387	45.5%
2	1,894	401	1,073	35.2
3	1,952	186	1,057	34.7
4	2,261	555	702	23.0
5	763	4,534	1,902	62.4

TABLE 12.2. Number of matches, erroneous matches, and erroneous non-matches using combinations of blocking criteria

Blocking criteria	Matched with correct parent (match)	Matched with wrong parent (erroneous match)	Not matched (erroneous non-match)	Percentage of erroneous non-matches
1	1,460	727	1,387	45.5%
1–2	2,495	1,109	460	15.1
1–3	2,908	1,233	112	3.7
1–4	2,991	1,494	39	1.3
1–5	3,007	5,857	22	0.7

that are in a fixed order or in fixed locations. Consequently criterion 4 is able to match fields of records having the following forms:

John K Smith

Smith J K Co

Criterion 3 (TELEPHONE) only produced 186 erroneous matches and only missed 1,057 actual matches. Criterion 5 (first 10 characters of NAME) had both the highest number of erroneous matches, 4,534, and the highest number of erroneous non-matches, 1,902.

Winkler [1984b] next considered combinations of these blocking criteria. His results are summarized in Table 12.2.

Criteria 1 and 2 are applicable to all EIA lists because all such lists have identified NAME and ADDRESS fields. As many lists obtained from sources external to EIA do not have telephone numbers, criterion 3 is not applicable to their records. As a number of EIA lists have consistently formatted NAME fields, criterion 4 will yield little, if any, incremental reductions in the number of erroneous non-matches during the blocking process. Finally, criterion 5 should not be used as a blocking criterion. Although the inclusion of criterion 5 reduces the number of erroneous non-matches from 39 to 22 and increases the number of correct matches by 16, it increases the number of erroneous matches by a staggering figure of 4,363 ($= 5,857-1,494$).

Results from using criteria 1 and 2 of Winkler's study are summarized in Table 12.3.

TABLE 12.3. Number of duplicate pairs identified using criteria 1 and 2

Criterion 1	Criterion 2		Total
	Present	Absent	
Present	859	601	1,460
Absent	1,035	?	
Total	1894		

Assuming that the necessary assumptions hold, we use the basic Lincoln–Peterson capture-recapture estimator of Section 6.3.1 to estimate the number of records in the blank cell and thereby estimate the total number of pairs of duplicates on the list of records:

$$\hat{N}_{LP} = \frac{x_{1+}x_{+1}}{x_{11}} = \frac{1460 \cdot 1894}{859} = 3,219.$$

This is reasonably close to the actual value of 3,050 duplicate records in the list.

12.4.   Using Blocking Strategies to Match Records Between Two Sample Surveys

In the next example, we use blocking strategies to efficiently match records between two record files. The larger file is the main file from the 2000 Decennial Census in the United States. This file has records on approximately 300 million individuals. The other file is the main Accuracy and Coverage Evaluation (ACE) file of approximately 750,000 individuals. The ACE is the 2000 version of the 1990 Post-Enumeration Survey (PES) but is twice as large as the 1990 PES. The 2000 Decennial Census file was constructed using optical scanning equipment to convert hand-written information to electronic form. The ACE data were collected via a Computer-Assisted Personal Interviewing (CAPI) interview. The ACE file is matched against the Census file by census blocks[2] in order to determine the extent of the overlap of the two files. The overlap was then used to estimate the undercount and over-count of the 2000 Decennial Census. The estimates that the Census Bureau produces by census blocks are ultimately used for reapportionment of Congressional districts and for Federal revenue-sharing.

Example 12.3: Using the 2000 ACE to evaluate alternative blocking strategies

After an exhaustive effort, the Census Bureau was able to determine that 606,411 records in the ACE file matched records within the 2000 Decennial Census file. This provided the Census Bureau (see Winkler [2004]) an opportunity to determine the efficacy of various blocking strategies. The results of using each of 11 blocking criteria, separately, are summarized in Table 12.4.

Except for Criterion 7, each criterion individually identified at least 70% of the matched pairs. Criterion 7 was designed to identify the matched pairs in which the first names and last names are switched. This accounted for 5.2% of the matched pairs.

The 11 blocking criteria listed above identified all but 1,350 matched pairs. The four best criteria – 1, 3, 11, and 9 – identified all but 2,766 matched pairs; while the five best criteria – 1, 3, 11, 9, and 8 – identified all but 1,966.

---

[2] Each census block consists of approximately 70 households.

TABLE 12.4. Number of matches identified by each blocking criterion

	Blocking criterion	Number of matches	Proportion of matches (in %)
1	Zip Code, First Character of Surname	546,648	90.1
2	First character of surname, First character of first name, Date of birth	424,972	70.1
3	10-digit telephone number	461,491	76.1
4	First three characters of surname, Area code of telephone number, House number	436,212	71.9
5	First three characters of first name, First three numbers of Zip Code, House Number	485,917	80.1
6	First three characters of surname, First three numbers of Zip Code, Area code of telephone number	471,691	77.8
7	First character of surname = First character of first name (2-way switch), First three digits of Zip Code, Area code of telephone number	31,691	5.2
8	First Three Digits of Zip Code, Day of Birth, Month of Birth	434,518	71.7
9	Zip Code, House Number	514,572	84.9
10	First three characters of surname, First three characters of first name, Month of Birth	448,073	73.9
11	First three characters of surname, First three characters of first name	522,584	86.2

Some of the matches that were not identified involved children living in a household headed by a single or separated mother. The children were listed under different last names in the two files. Their date of birth was missing in the ACE file and the street address of their residence was missing in the Census file. The records on these children also had names with a high rate of typographical error that may be at least partially due to scanning error.[3] Table 12.5 illustrates the nature of these typographical problems.

TABLE 12.5. Examples of typographical problems that cause matches to be missed (artificial data)

Relationship to head of household	Record in census file		Record in ACE file	
	First name	Last name	First name	Last name
Head of Household	Julia	Smoth	Julia	Smith
Child #1	Jerome	Jones	Gerone	Smith
Child #2	Shyline	Jones	Shayleene	Smith
Child #3	Chrstal	Jones	Magret	Smith

---

[3] In 2000, names were captured (i.e., entered into computer files) via a procedure that scanned hand-written text and converted it to characters. Prior to 2000, names were not captured in Decennial Censuses.

The names on the childrens' records that should be matched have no 3-grams in common. Here, we say two names have a 3-gram in common if any three consecutive letters from the name on one record appear in the corresponding name on the other record. Consequently, it is unlikely that such matched pairs could be linked through any computerized procedure that uses only the information in the Census and ACE files.

12.5.   Estimating the Number of Matches Missed

If an estimate of the number of matches missed when comparing two lists is required, then the lists can be sampled directly. However, even if very large samples are selected, Deming and Gleser [1959] show that the estimated variances of the error rate (i.e., the proportion, p, of missed matches) often exceed the values of the estimates. This is particularly true when the error rate, p, is small.

If samples are not used, then capture–recapture techniques can be used to estimate the number of matches missed, as illustrated in the example of Section 12.3. Winkler [2004; Section 4] uses sophisticated (capture–recapture) log-linear models to estimate the number of matches missed in the problem considered in Example 12.3. Based on the capture–recapture estimates, Winkler estimated that no more than 1 in 10^{11} pairs of the remaining 10^{17} (300 million $\times$ 300 million) pairs that were not in the entire set of pairs obtained via blocking would be a match.

12.6.   Where Are We Now?

In this chapter, we showed how blocking strategies could be used in record linkages to limit the number of pairs that need to be compared while at the same time minimizing the number of matches that are missed. In the next chapter, we discuss string comparator metrics. These can be used to account for minor data-entry errors and thereby facilitate record linkages. The topic of string comparator metrics is the last enhancement to record linkage that we present.

13
String Comparator Metrics for Typographical Error

Many fields such as the first name and last name occasionally contain minor typographical variations or errors. Even with high-quality lists, such as the 1990 US Census and itsPES, there were areas of the United States in which 30% of the first names and 25% of the last names of individuals who were in fact matches did not agree exactly on a character-by-character basis. If we attempt to match two such records and at least one of the first names on the two records has a typographical variation, then we may fail to match two records that may indeed be matches. Using the terminology of computer science, we can consider these records or their components to be *strings* – that is, strings of alphanumeric characters. We need a practical method for dealing with such situations.

As the name indicates, string comparator metrics are used to compare two strings. In particular, they are used to determine how much alike the two strings are to each other. Common practice is to restrict the values of the metrics to the interval from zero to one; here, one indicates perfect agreement (the two strings are identical) and zero indicates that they are highly dissimilar, the extreme case being that they have no characters in common. These values are needed to adjust the likelihood ratios of the Fellegi–Sunter scheme to account for this partial agreement. In this work we focus on the string comparator metric introduced by Jaro and enhanced by Winkler. Current research in this area is described in Cohen, Ravikumar, and Fienberg [2003a, b].

13.1. Jaro String Comparator Metric for Typographical Error

Jaro [1972, see also 1989] introduced a string comparator metric that gives values of partial disagreement between two strings. This metric accounts for the lengths of the two strings and partially accounts for the types of errors – insertions, omissions, or transpositions – that human beings typically make when constructing alphanumeric strings. By *transposition* we mean that a character from one string is in a different position on the other string. For example, in comparing "sieve" to "seive," we note that "i" and "e" are transposed from one string to the other. The string comparator metric also accounts for the number of characters the two strings have in common. The definition of *common* requires that the agreeing characters must be within half of the length of the shorter string. For example, "spike" and "pikes" would only have four characters in common because the "s's" are too far apart.

Specifically, let s_1 denote the first string, s_2 denote the second string, and let c denote the number of characters that these two strings have in common. Then, if $c > 0$, the Jaro string comparator metric is

$$\Phi_J(s_1, s_2) = W_1 \cdot \frac{c}{L_1} + W_2 \cdot \frac{c}{L_2} + W_t \cdot \frac{(c - \tau)}{c}$$

where

W_1 is the weight assigned to the first string,
W_2 is the weight assigned to the second string,
W_t is the weight assigned to the transpositions,
c is the number of characters that the two strings have in common,
L_1 is the length of the first string,
L_2 is the length of the second string, and
τ is the number of characters that are transposed.
We require that the weights sum to 1: $W_1 + W_2 + W_t = 1$.
Finally, if $c = 0$, then $\Phi_J(s_1, s_2) = 0$.

Example 13.1: Using the Jaro string comparator metric on Higvee versus Higbee

Let the first string, s_1, be "Higbee" and the second string, s_2, be "Higvee," and let all of the weights be equal to 1/3. Find the value of the Jaro string comparator metric for these two strings.

Solution

Because the two have five of six letters each in common, we have

$$L_1 = L_2 = 6, \ c = 5, \ \text{and} \ \tau = 0.$$

Hence,

$$\Phi_J(s_1, s_2) = W_1 \cdot \frac{c}{L_1} + W_2 \cdot \frac{c}{L_2} + W_t \cdot \frac{(c - \tau)}{c} = \left(\frac{1}{3}\right) \cdot \left(\frac{5}{6}\right) + \left(\frac{1}{3}\right) \cdot \left(\frac{5}{6}\right)$$
$$+ \left(\frac{1}{3}\right) \cdot \left(\frac{5 - 0}{5}\right) = \frac{8}{9}.$$

Example 13.2: Using the Jaro string comparator metric on Shackleford versus Shackelford

Let the first string, s_1, be "Shackleford" and the second string, s_2, be "Shackelford" and let all of the weights be equal to 1/3. Find the value of the Jaro string comparator metric for these two strings.

Solution

Because two letters "l" and "e" are transposed, we have

$$L_1 = L_2 = 11, \ c = 11, \ \text{and} \ \tau = 2.$$

Hence,

$$\Phi_J(s_1, s_2) = W_1 \cdot \frac{c}{L_1} + W_2 \cdot \frac{c}{L_2} + W_t \cdot \frac{(c - \tau)}{c} = \left(\frac{1}{3}\right) \cdot \left(\frac{11}{11}\right) + \left(\frac{1}{3}\right) \cdot \left(\frac{11}{11}\right)$$
$$+ \left(\frac{1}{3}\right) \cdot \left(\frac{11 - 2}{11}\right) = \frac{31}{33}.$$

13.2. Adjusting the Matching Weight for the Jaro String Comparator

To take into account the partial weight from the Jaro string comparator metric, Winkler [1990] suggested redefining the weights when there is agreement, as the adjusted weight

$$w_{am} = \begin{cases} w_a & \text{if } \Phi_J = 1 \\ \max\left[w_a - \{(w_a - w_d) \cdot (1 - \Phi_J) \cdot (4.5)\}, w_d\right] & \text{if } 0 \le \Phi_J < 1 \end{cases}$$

where *the full agreement weight* is $w_a = \log_2\left(\frac{m}{u}\right)$ and the *full disagreement weight* is $w_d = \log_2\left(\frac{1-m}{1-u}\right)$. The constant 4.5 controls how quickly decreases in partial agreement factors force the adjusted weight to the full agreement weight. We can write the full agreement weight, w_a, for example, as

$$w_a = \log_2\left(\frac{m}{u}\right) = \log_2\left(\frac{P[\Phi_J = 1 \mid r \in M]}{P[\Phi_J = 1 \mid r \in U]}\right).$$

13.3. Winkler String Comparator Metric for Typographical Error

Winkler [1990] introduced the following enhanced version of the Jaro string comparator metric:

$$\Phi_W(s_1, s_2) = \Phi_J(s_1, s_2) + i \cdot 0.1 \cdot (1 - \Phi_J(s_1, s_2))$$

where $i = \min(j, 4)$ and j, in turn, is the number of initial characters the two strings have in common on a character-by-character basis. This metric gives more importance to agreement on the initial characters of the strings than to agreement on the later characters of the string. This idea was inspired by the results of a large empirical study that Pollock and Zamora [1984] conducted to develop a spell checker. (This was part of a Chemical Abstracts Service project funded by the US National Science Foundation.) Their study concluded that the most reliable characters of a string are those that occur first and that the data quality deteriorates monotonically as one moves from the beginning of the string to the end of the string. Winkler's enhancement increases the metric by a constant amount, $1 - \Phi_J(s_1, s_2)$, for each of the consecutive initial characters that match exactly between the two strings, up to a maximum of four characters.

Example 13.3: Using the Winkler string comparator metric on Shackleford versus Shackelford

Compute the Winkler metric under the conditions of Example 13.2.

Solution

Here, we have the first five letters of the two strings in character-by-character agreement. Hence, $i = 5$ and $j = \min(i, 4) = \min(5, 4) = 4$. So,

$$\Phi_W(s_1, s_2) = \Phi_J(s_1, s_2) + i \cdot 0.1 \cdot (1 - \Phi_J(s_1, s_2)) = \frac{31}{33}$$

$$+ 4 \cdot 0.1 \cdot \left(1 - \left(\frac{31}{33}\right)\right) = 0.9636.$$

13.4. Adjusting the Weights for the Winkler Comparator Metric

Instead of applying identical adjustment schemes to each of the diverse fields (e.g., last name, first name, and street number), Winkler and Thibaudeau [1991] proposed a scheme under which each of the fields would be assigned a distinct, but more appropriate, weight. This scheme involved the use of a piecewise linear function and required representative sets of pairs for which the truth of matches is known. The newly adjusted weights are specified as:

$$w_{na} = \begin{cases} w_a & \text{if } \Phi_W \geq b_1 \\ \max[\{w_a - (w_a - w_d) \cdot (1 - \Phi_W) \cdot a_1\}, w_d] & \text{if } b_2 \leq \Phi_W < b_1 \\ \max[\{w_a - (w_a - w_d) \cdot (1 - \Phi_W) \cdot a_2\}, w_d] & \text{if } \Phi_W < b_2 \end{cases}$$

The constants a_1, a_2, b_1, and b_2 depend on the specific type of field (e.g., last name) to which the weight adjustment is being applied. In most practical applications, we have $a_1 < a_2$. Some of the specific constants used are given in Table 13.1.

While this individual-adjustment scheme led to a slightly more accurate matching scheme with the high-quality files of the 1990 Census, it usually hurts matching with other files that are of lesser quality. For general matching applications, the Census Bureau applies the Jaro–Winkler scheme of Section 13.3 to the surname when it is used as a matching variable.

TABLE 13.1. Constants used in piecewise linear weight adjustments

Field type	Constant			
	a_1	a_2	b_1	b_2
Given name	1.5	3.0	.92	.75
Surname	3.0	4.5	.96	.75
House number	4.5	7.5	.98	.83

13.5. Where are We Now?

We have now completed Part II, and thereby completed all of our methodology chapters on editing, imputation, and record linkage. All of this is summarized in Chapter 20, our Summary chapter. In Chapters 14–17 we present case studies illustrating the application of these methodologies. In particular, Chapter 14 describes an application of both basic editing techniques and record linkage procedures to detect and repair errors in a large database.

Part 3
Record Linkage Case Studies

Introduction to Part Three

We devote Part 3 of our book to discussing real world examples of record linkage techniques that have been used in a wide range of applications. Chapter 14 begins this discussion with examples that we have personally experienced at HUD with respect to FHA-insured single-family mortgages. Chapter 15 reviews record linkage experiences in other areas of interest with case studies on medical, biomedical, and highway safety applications. Chapter 16 discusses address list frame and administrative list construction, using examples from Statistics Canada, US Department of Agriculture, and US Census Bureau experience.

14
Duplicate FHA Single-Family Mortgage Records

A Case Study of Data Problems, Consequences, and Corrective Steps[1]

14.1. Introduction

We begin our Part III presentation of case studies of data quality/record linkage activities with our own experience detecting and correcting errors in a very large database of FHA single-family mortgage records. All of the examples we discuss in this chapter have been modified to protect individual privacy and/or simplify the exposition. In order to illustrate the techniques discussed in Chapter 5, we focus on identifying and repairing errors on records of FHA loans originated prior to the implementation of a new system that had a large number of front-end edits. These front-end edits have been successful in preventing erroneous data from entering certain fields on these mortgage records.

14.1.1. Background

The Federal Housing Administration (FHA), an agency within the US Department of Housing and Urban Development (HUD), insures mortgages against the risk that the borrower, for whatever reason, will be unable to continue making payments on his/her mortgage. The FHA's mortgage guarantee insurance programs are partitioned into four separate insurance funds. FHA's Mutual Mortgage Insurance Fund (MMIF) is its largest fund. The MMIF insures mortgages on single-family homes consisting primarily of single-family detached houses and townhouses. FHA reported that as of July 1, 2005, the MMIF had about 4 million mortgages insured with an aggregate face amount of about $375 billion. FHA's General Insurance Fund (GIF) also insures mortgages on single-family homes. In addition, the GIF insures loans on individual condominium units as well as on apartment buildings, nursing homes, hospitals, and mobile homes. The Cooperative

---

[1] An earlier version of this chapter was coauthored with William J. Eilerman, Office of Evaluation, US Department of Housing and Urban Development.

Management Housing Insurance Fund (CMHIF) insures mortgages on cooperative apartment buildings. The Special Risk Insurance Fund (SRIF) insures mortgages on single-family homes, excluding condominiums, and apartment buildings.

Since its inception in 1934, the FHA has insured over 32 million mortgages on single-family homes. The bulk of these mortgages have been insured under FHA's MMIF. The mutuality feature of this fund means that dividends (also known as distributive shares) may be paid to certain borrowers when they terminate their insurance. The amount of the dividend depends on the mortgage amount, the year the mortgage began amortizing, and the amortization plan of the mortgage. On September 1, 1983, FHA instituted a one-time premium collection system whereby the entire premium is paid in advance, and unearned premium refunds are paid to those who successfully terminate their loans prior to maturity.

14.1.2. The FHA Single-Family Data Warehouse

Currently, FHA's primary single-family database is known as the FHA single-family data warehouse. While the vast majority of its records do not have serious data problems, we have identified some with severe problems. Because of the enormous size of the database, an error rate of only 1 or 2% can result in hundreds of thousands of incorrect records. Our intent here is to describe some of the data problems we have investigated as well as the efforts we have made to improve the accuracy of the affected records.

14.1.3. The Purpose and Problems of the Data Warehouse

The FHA staff as well as other analysts use the FHA single-family data warehouse to answer basic questions about the FHA single-family insurance programs. These include the number of mortgages insured, the number of mortgage insurance claims, and the number of prepayments. The data are also used as input by researchers constructing actuarial/statistical/econometric models to predict (1) the number/dollar amount of future insurance claims, (2) the number/dollar amount of future prepayments, and (3) the economic value of each of the FHA single-family insurance funds. Unfortunately, some mortgages have been entered into FHA's single-family data warehouse under two or more identification numbers[2], usually with only slight differences between the identification numbers and few, if any, differences in the case data. When such records are displayed together, it is usually apparent that both represent the same mortgage; however, because the identification numbers are different, the data warehouse treats both records as unique, individual mortgages.

A much more pervasive problem is one of incorrect termination statuses. When a borrower refinances or prepays an FHA-insured single-family mortgage, the lender servicing that mortgage is supposed to so notify FHA, and FHA is

---

[2] These identification numbers are known as FHA case numbers and are discussed in detail in Section 14.2.

supposed to make the appropriate changes to its databases. But this process does not always work as intended. As a consequence, FHA has hundreds of thousands of mortgage records in its single-family data warehouse with a termination status of "active" when, in fact, the underlying mortgage has been refinanced, paid in full, or terminated for some other reason. Because (1) FHA only insures "first" mortgages, that is, those with a primary lien on the underlying property and (2) no condominium units are supposed to be insured under the MMIF, it follows that there should be at most one "active" MMIF-insured mortgage per property address. Too often, this is not the case.

In the next section, we discuss the concept of an FHA case number in order to facilitate the rest of the discussion. In Section 14.3, we discuss the problem of the duplicate mortgage records. In Section 14.4, we discuss the problem of the mortgage records with the incorrect termination statuses. In both sections, we begin by illustrating the nature of the problem; we then describe the procedures we used to identify and correct these problems and list some of the consequences of these errors. Finally, in Section 14.5, we describe a scheme for estimating the number of duplicate mortgages records.

14.2.   FHA Case Numbers on Single-Family Mortgages

Beginning in January, 1962, all FHA case numbers consisted of a 3-digit State and Office code prefix (SSO-) followed by a dash, a six-digit serial number, and a final check digit (SSO-DDDDDDC). The use of the check digit was intended to improve the accuracy of the FHA case number; unfortunately, it was easy to circumvent the protective function of the check digit by appending an "X" to the end of any questionable case number. This causes one of HUD's computer systems to calculate and insert the appropriate check digit for any number entered, thus passing over the incorrect or missing check digit. This, in turn, allows processing of the invalid FHA case to continue and even gives its case number an aura of legitimacy, because it has a "correct" check digit attached from that point on. Fortunately, this problem was eliminated with the implementation of HUD's Computerized Homes Underwriting Management System (CHUMS) in the mid-1980s, as CHUMS assigns FHA case numbers automatically, leaving no chance for manual error.

14.3.   Duplicate Mortgage Records

The first problem we considered was the existence of thousands of mortgage records entered into HUD's single-family data warehouse under two or more FHA case numbers.

14.3.1.   Examples of Problem Cases Found in the Data Warehouse

We now discuss some representative examples of the problems discussed above:

Example 14.1:

Consider the following case record:

FHA Case Number	Street Address	City	State	Zip Code	Name(s) of Borrower
441-1451573	714 Grand Ave	Cincinnati	PA	45232	Hollins, Larry & Juanita

Here we have a record with a Philadelphia, Pennsylvania, office code of "441" but a Cincinnati, Ohio Zip Code. Further research showed that (1) the office code should have been entered as "411" and (2) the data warehouse also had a record for FHA case number 411-145157x with essentially the same information as the record for case number 441-1451573.

Example 14.2:

Consider the following case record:

FHA Case Number	Street Address	City	State	Zip Code	Name(s) of Borrower
441-2760279	1309 Broad	Richmond	PA	23231	Brown, Robert

Here we have a record with a Philadelphia, Pennsylvania, office code of "441" but a Richmond, Virginia Zip Code. Further research showed that (1) the office code should have been entered as "541" and (2) the data warehouse also had a record for FHA case number 541-276027x with essentially the same information as the record for case number 441-2760279.

Both of these examples are instances in which internal consistency checks (see Chapter 5) were used to identify erroneous entries within a mortgage record that led to the detection of a duplicate mortgage record.

Example 14.3:

We next consider the following six contiguous case records extracted from FHA's data warehouse in ascending order of their FHA case numbers:

FHA Case Number	Initial Mortgage Amount (in dollars)	Begin- Amortization- Date	Contract Interest Rate (in %)	Name(s) of Borrower
131-5132066	70,100	Aug-1987	10.5	Copp, Jeffrey S.
131-5132095	47,750	Oct-1987	11.0	Budge, Robert W.
131-5132116	76,500	Aug-1987	10.5	Flood, Sherrie D.
131-5132151	68,800	Jun-1984	14.5	O'Meara Tom L Kim A
131-5132180	47,600	Aug-1987	10.5	Els, Sherri A.
131-5132197	43,250	Oct-1987	10.5	Hollins, Richard M.

In the third column of the table, we observe that the fourth record appears to be out of sequence. The year this loan began to amortize was 1984, versus 1987 for the five other loans shown. Moreover, its contract interest rate of 14.5% is much higher than those of the other five loans. Further examination revealed that the correct FHA case number for this mortgage was 131-3807331; this record was a duplicate that needed to be deleted from the FHA single-family data warehouse. In a sense then, the record on FHA case number 131-5132151 has an inconsistency between its case number and both (1) its begin-amortization-date and (2) its interest rate.

Example 14.4:

In this example, we examine, also in ascending order of the FHA case number, the first six case records on the FHA single-family data warehouse within office "372" having begun amortization date on or after January 1, 1975.

FHA case number	Property address	Begin-amortization-date	Contract interest rate (in %)	Name(s) of borrower
372-0101064	395 Kings Highway Amherst, NY	Dec-1979	11.5	Spruce, Ronald H
372-0113867	360 Rand St Rochester, NY 14615	Apr-1983	12.0	Stewart P D Q
372-0116726	338 Spencer Road Rochester, NY 14609	Jun-1983	12.0	Landrieu M J
372-0707605	356 Hazelwood Ave Buffalo, NY 14215	Jan-1975	9.5	Richardson BF L
372-0707736	352 Ackerman St Rochester, NY 14609	Jan-1975	9.0	Bowers Newton P
372-0708494	322 Worcester Place Buffalo, NY 14215	Jan-1975	9.0	McDonald A L

We note here that the first three records listed appear to be out of sequence – or, using the terminology of Chapter 5 – are out of range. Further research revealed that the FHA case numbers corresponding to these three records should have been 372-101064x, 372-113867x, and 372-116726x, respectively. This is an example of what Chapter 5 calls a range test.

The approaches just described were relatively naive. With the assistance of the staff in HUD's Office of Information Policy Systems, we developed a more sophisticated approach to detecting problems. The idea was to use automated record linkage techniques to match cases. An exact-match scheme we ran successfully involved finding records with identical serial numbers and identical mortgage amounts. At the time this work was done initially, the database used had around 10 million records. The result was a file consisting of nearly 200,000 putative matches from which we identified about 5,000 duplicate records by doing additional programming in the APL computer language. We consider two examples of this work.

Example 14.5:

FHA Case Number	Property Address	Initial Mortgage Amount (in $)	Begin-Amortization-Date	Interest Rate (in %)	Name(s) of Borrower
372-1132617	109 Elm St Bethany, NY 14054	33,000	Feb-1983	12.5	Brown Thomas N
374-1132613	109 Elm St Bethany, NY 14054	33,000	Feb-1983	12.5	Brown TR

Here, the two records match on the serial portion of their FHA case numbers – 113261 – as well as on their initial mortgage amounts –$33, 000. It is clear that they also match on their begin-amortization-dates, interest rates, property addresses, and borrower name(s).

Example 14.6:

FHA Case Number	Property Address	Initial Mortgage Amount (in $)	Interest Rate (in %)	Name(s) of Borrower
371-0912411	3111 Stratford Ave Bronx, NY 10464	54,950	15.5	Bryan William Randolph
374-0912416	3111 Stratford Ave Bronx, NY 10464	54,950	15.5	Bryan William Randolph

Here again, the two records match on the serial portion of their FHA case numbers – 091241 – as well as on their initial mortgage amounts –$54, 950.

14.3.2.   Procedures for Identifying Duplicate Mortgage Records – Where Are We Now?

We just described several schemes used to identify duplicate case records within FHA's single-family data warehouse. One scheme identified cases whose FHA case number was not consistent with the Zip Code on its property address. This worked well on cases having a Zip Code and turned up about 1,000 duplicates. Unfortunately, it was of only limited use because millions of case records do not have a Zip Code in the data warehouse. Another scheme (see Example 14.4) focused on cases with the highest serial numbers and the lowest serial numbers for each office.

Questionable records were investigated individually (i.e., manually) using an information-retrieval system. Such records were then matched with other records based on two or three key data elements.

14.3.3.   Some Consequences of Duplicate Mortgage Records

Among the bad consequences of the errors described above, the following are particularly noteworthy:

- Valid FHA cases may not be entered onto the single-family data warehouse because another case has previously been entered onto the system with the same FHA case number.
- Claim payment requests may be delayed because the data on the case cannot be found on the single-family data warehouse under the correct FHA case number. This will occur, for example, if the FHA case number is not entered onto the system correctly. Such delays can be expensive as they increase HUD's interest costs.
- Problems with unearned premium refunds are likely. For example, HUD could pay both an insurance claim and an unearned premium refund on the same FHA-insured mortgage, or HUD could pay two or more unearned premium refunds on the same mortgage.
- Such problems can lead to fraud as unscrupulous tracers pressure borrowers to accept multiple unearned premium refund payments, or as HUD employees or contractors attempt to take advantage of the situation.
- Such errors also contribute to the unfavorable publicity HUD receives for not finding borrowers who are supposedly owed money by HUD. This is in contrast to the vigorous efforts other US government agencies (e.g., the Internal Revenue Service) make to collect money owed the US government.
- Such errors have a deleterious effect on the financial condition of the MMIF. Even when a check is not sent to a borrower for a refund payment, the fund is nevertheless debited when an unearned premium is declared on a mortgage record in the data warehouse.
- Finally, such errors have a deleterious effect on these important databases and adversely affect their use for analytical (e.g., statistical and actuarial) studies. Even when the data are 100% accurate, it is a complex problem to construct accurate claim and prepayment models for the MMIF.

14.4.   Mortgage Records with an Incorrect Termination Status

The existence of hundreds of thousands of mortgage records residing on the single-family data warehouse and having a termination status of "active," when the underlying mortgage has actually terminated, usually by prepayment, is the second major problem we discuss in this chapter.

14.4.1. Examples of Mortgages with Incorrect Termination Status

Example 14.7:

FHA Case Number	Property Address	Begin-Amortization-Date	Status of Loan	Interest Rate (in %)	Name(s) of Borrower
371-1019310	109-07 211th Place Queens, NY 11429	Mar-1982	Active	16.5	Smith John Paulette
374-4413730	109-07 211th Place Queens, NY 11429	Sep-2004	Active	5.5	Smith, Paulette

In this instance, the data warehouse lists two active mortgages on a property located in Queens Village, New York, where in reality the first mortgage that was originated in March of 1982 at an interest rate of 16.5% has been refinanced at least once, that is, during September 2004.

Example 14.8:

FHA Case Number	Property Address	Begin-Amortization-Date	Status of Loan	Interest Rate (in %)	Name(s) of Borrower
372-1221854	323 OAK ST BUFFALO, NY 14215	Apr-1984	Active	13	J. & K. Falkides
372-1519223	323 OAK ST BUFFALO, NY 14215	Jan-1987	Prepaid	9	Falkides, John P

In this instance, the data warehouse lists one active mortgage and a later mortgage terminated by prepayment for a property in Buffalo, New York. It appears that the first mortgage was originated during April of 1984 at an interest rate of 13% and refinanced during January of 1987 at an interest rate of 9%.

Example 14.9:

FHA Case Number	Property Address	Begin-Amortization-Date	Status of Loan	Interest Rate (in %)	Name(s) of Borrower
131-5109339	3104 RTE 1750 COAL VALLEY, IL 61240	Oct-1986	Active	9.0	Angel, Dale J
131-7200510	3104 RTE 1750 COAL VALLEY, IL 61240	Jul-1994	Claimed	7.5	Angel, Dale E

In this instance, the data warehouse lists one active mortgage and a later mortgage terminated by insurance claim for a property in Coal Valley, Illinois. Again, it appears that the first mortgage was originated during October of 1986 at an interest rate of 9% and refinanced during July of 1994 at an interest rate of 7.5%. The second loan eventually resulted in an insurance claim being paid by HUD.

14.4.2. Procedures Used to Identify Mortgage Records with Incorrect Termination Status

Here again, we used an exact-match scheme. We first blocked the data based on the first four digits of the Zip Code of the property address. So, we excluded all records that lacked a Zip Code. For each remaining record, we created a string consisting of 18 characters. The first part of the string consisted of the first 10 alphanumeric characters of the street address of the insured property. The next four characters of the string were the first four alphabetic characters of the name of the borrower. The next two characters represented the month the loan began amortizing while the last two characters represented the last two digits of the year the long began amortizing.

We then identified pairs of mortgage records having duplicate strings. We considered these to be duplicate records and deleted the erroneous records from our database.

We then deleted the last four characters from the remaining strings and again identified pairs of mortgage records having duplicate strings. The hope was that these represented pairs of mortgages with identical property addresses and borrowers. We considered most of these to represent FHA mortgages that refinanced into new FHA mortgages. If the loan record with the earlier begin-amortization-date was still listed as being an "active" loan, then we changed the termination status of that loan record to "termination by prepayment." In most instances, we added the begin-amortization-date of the later loan as the termination date of the earlier one.

Finally, we deleted the last four characters of the rest of the string, leaving us with only the 10 characters from the property address field. We again identified pairs of mortgage records having duplicate strings. The hope was that these represented pairs of mortgages with identical property addresses. We considered many of these to represent FHA mortgages on houses that were sold and where the new homebuyer also used FHA insurance. These pairs of records required staff review. If the loan record with the earlier begin-amortization-date was still listed as being an "active" loan and other criteria were met as well, then we changed the termination status of that loan record to "termination by prepayment." In most instances, we added the begin-amortization-date of the later loan as the termination date of the earlier one.

Thus far, this process has worked reasonably well in that it has enabled us to identify thousands of duplicate records and helped us to correct the termination status of tens of thousands of other records as well. Future plans call for us to use probabilistic record linkage techniques. However, exact matching techniques are currently giving HUD staff more than enough records to consider.

14.4.3. Some Consequences of Incorrect Termination Status

Among the bad consequences of these errors in termination status, the following stand out:

- The amounts of insurance in force are all overstated for the MMIF, the GIF, and the SRIF.
- Lenders are paying periodic mortgage insurance premiums on mortgages that have already terminated.
- Borrowers have not been paid unearned premium refunds or distributive shares to which they are (or were at one time) entitled.
- Such errors have a deleterious effect on these important databases and adversely affect their use for analytical (e.g., statistical and actuarial) studies. As with the previous problem of duplicate mortgage records, incorrect termination status makes it impossible, rather than merely complex, to construct accurate claim and prepayment models for the MMIF.

14.5. Estimating the Number of Duplicate Mortgage Records

We would like to use the results of Sections 14.3.2 and 14.4.2 to estimate the total number of duplicate records in FHA's single-family data warehouse. Unfortunately, we cannot do this for a variety of reasons. One reason is that the data were analyzed at different times, actually several years apart. Another is that it would require us to determine which duplicate records had been identified by both schemes. This was, of course, not possible because shortly after we identified the records as duplicates we deleted them from the data warehouse. Nevertheless, to illustrate the methodology, let us suppose that they had been analyzed at roughly the same time, that we had been able to determine exactly the number of records identified by both schemes, and that the data were as summarized in Table 14.1.

TABLE 14.1. Number of duplicate records

Method of Section 14.4.2	Duplicate Record	Method of Section 14.3.2 Duplicate Record	
		Yes	No
	Yes	2,000	500
	No	400	??

Using the estimator for $\hat{N}$ given in Section 6.3.1 as the estimate of the total number of duplicate records, we would obtain

$$\hat{N} = \frac{(2,000 + 500) \cdot (2,000 + 400)}{2,000} = 3,000.$$

15
Record Linkage Case Studies in the Medical, Biomedical, and Highway Safety Areas

15.1. Biomedical and Genetic Research Studies

Perhaps the earliest use of the computer to carry out record linkage is that described by Newcombe, Kennedy, Axford, and James [1959]. These researchers were interested in two possible applications of record linkage:

1. To keep track of large groups of individuals who had been exposed to low levels of radiation in order to determine the causes of their eventual deaths.
2. To assess the relative importance of (i) repeated natural mutations and (ii) fertility differentials, in maintaining the frequency of genetic defects in the human population.

They felt that an extensive history of an individual could be constructed by matching the vital records on that individual maintained by the various governmental agencies. At the time their paper was published, Newcombe, Kennedy, and James were affiliated with Atomic Energy of Canada, Ltd, located in Chalk River, Ontario, and Axford with the Health and Welfare Division of the Dominion Bureau of Statistics in Ottawa. Their paper deals primarily with the second application listed above.

Their first step was to develop a method for linking birth records and marriage records. These researchers considered the records on (1) the 34,138 births that occurred in the Canadian province of British Columbia during the year 1955 and (2) the 114,471 marriages that took place in British Columbia during the years 1946–1955. They carried out an intensive study of the various sources of errors on the record linkage procedure for about 20% of such records.

In order to reduce the number of marriage records within the matching process, the researchers decided to consider both the husband's surname and the wife's maiden name. They anticipated that only rarely would the number of matched pairs of surnames exceed one or two per birth. The researchers used the Soundex coding system to encode both of these surnames. They hoped that this system would help them avoid problems with common spelling variations as well as to bring together linkable records that would have been widely separated if arranged in a strictly alphabetic sequence.

To further distinguish between married couples, the researchers relied primarily on the following identifying items (in addition to the Soundex-coded surnames):

1. Full alphabetic surnames of the husband and wife (limited to nine letters each)
2. Their provinces or countries of birth (each coded as a two-digit number)
3. The first initial of each of their given names.

The researchers had the following items available as well:

- The ages of the married couple, which were available on the birth records and, for those married between 1951 and 1956, on the marriage record as well
- Their middle initials (via the birth record)
- The name of the city where the marriage took place
- The name of the city where the baby was born.

Most of the time the researchers had little difficulty matching the names from the marriage record to the names on the birth records. However, this did prove difficult for them occasionally. In such cases they constructed an algorithm to compute the log odds of agreement versus disagreement. The following were the variables used as input to this algorithm:

- The two surnames
- The two birthplaces
- The two first initials
- The two ages (where available)
- The place of the event (i.e., the city of the birth or marriage).

Specifically, the log odds (to the base 2) of linkage or non-linkage were obtained from

$$\log_2 p_L - \log_2 p_U = \log_2 \left(\frac{p_L}{p_U} \right)$$

where p_L is the relative frequency of the observed outcome among linked pairs of records and p_U is the relative frequency of the observed outcome among unlinked pairs – that is, pairs of records that have been brought together by chance. Note that the log odds will be positive when the records agree and will be negative when they do not agree. For example, if we are comparing the first initial on an individual, we have two possible outcomes:

(1) Agreement – Here the frequency ratio is

$$\frac{p_L}{p_U} = \frac{\text{frequency of AGREEMENT among linked pairs}}{\text{frequency of AGREEMENT among unlinked pairs}}.$$

(2) Disagreement – Here the frequency ratio is

$$\frac{p_L}{p_U} = \frac{\text{frequency of DISAGREEMENT among linked pairs}}{\text{frequency of DISAGREEMENT among unlinked pairs}}.$$

The researchers estimated (1) that they obtained an accuracy rate of "about 98.3 percent of the potential linkages" and (2) "that contamination with spurious linkages is 0.7 percent." They concluded that "It is doubtful whether the present accuracy of the procedure can be matched by that of conventional survey and interview techniques, and its potential accuracy is certainly much greater than that of conventional techniques." One might also surmise that the cost was just a small fraction of that of a conventional household survey.

15.2.   Who goes to a Chiropractor?

15.2.1.   Purpose of Study

The purpose of this study (see Nitz and Kim [1997]) was to link insurance company records with police motor vehicle accident reports in order to answer the following questions:

- Who goes to a chiropractor?
- What is the relationship between occupant, vehicle accident and crash characteristics and the choice of care?

15.2.2.   Background

Hawaii requires resident owners of automobiles to have no-fault automobile insurance. In the event that someone is injured in an automotive accident in Hawaii, the victim has the option of choosing treatment from a number of types of medical providers. In addition to MDs, such providers might include chiropractors, physical therapists, and massage therapists. The two types most frequently chosen are medical doctors (MDs) and doctors of chiropractic (DCs).

15.2.3.   Data

A police motor vehicle accident report is required in Hawaii on all accidents on public roads that involve personal injury, death, or estimated property damages above a certain threshold – for example, $1,000 in 1990. The reporting form contains a description of the accident, the date and time of the accident, the severity of injury as well as the birth date and gender of the driver.

An insurance company's accident file consisted of 6,625 closed case records for accidents occurring in Hawaii during calendar years 1990 and 1991. Drivers as well as injured passengers and pedestrians were identified by their birth date and gender as well as by the date and time of the accident.

A separate insurance payment file contained payment information on about 58,000 transactions involving these 6,625 claims. This payment file was used to obtain the type of medical provider(s) selected.

15.2.4.   *Matching Process*

In order to conduct the analysis, it was necessary to link the automobile insurer's accident and payment files for each injured person to the corresponding police motor vehicle accident report file. The driver's birth date and gender as well as the date and time of the accident were the fields used to make this linkage.

Matchware Technology's Automatch 4.0 was the software used. Three passes were made. The first pass partitioned the files into homogeneous blocks based on the driver's age and sex, and matched on the date of the accident, the time of the accident, and the birth date of the driver. This produced 2,565 matches. The second pass blocked on the date of the accident and the sex of the driver, and matched on the time of the accident and the driver's birth date. This produced an additional 1,001 matches. A third and final pass was designed to pick up cases that were not matched due to errors in the recorded time of the accident. This produced an additional seven matches, bringing the total number of matched records up to 3,573. In short, the researchers were able to create usable records on 3,573 people injured in automobile accidents during 1990 and 1991.

15.2.5.   *Results*

The following were the main findings of this study:

- Only 7% of those injured in automobile accidents that sought treatment from either a medical doctor or a chiropractor went exclusively to a chiropractor.
- Of those injured in automobile accidents, 65.1% in the 21–34 age group chose to be treated by a chiropractor whereas only 56.0% in the other age groups sought chiropractic care.

15.3.   National Master Patient Index

According to Bell and Sethi [2001], the National Institute of Science and Technology (NIST) has funded a study (NIST Advanced Technology Grant 97-03) "to help accelerate the development of a massively distributed" National

Master Patient Index (MPI). The idea behind this is to create a complete, computerized medical history of all individual medical patients in the United States. The authors list four advantages of an MPI:

1. Elimination of redundant medical tests and procedures
2. Speedier access to a patient's medical history in the event of a medical emergency
3. Elimination of the need to re-obtain a patient's medical history
4. A more accurate medical history.

Some disadvantages of an MPI are:

• Concerns about the confidentiality of sensitive medical information. Will this inhibit individuals from seeking help from a mental health professional?
• The difficulty of avoiding errors, particularly mismatches of individuals. How many individuals in the United States, for example, are named Mary Smith or Alan Jones?

In spite of these disadvantages, the paper of Bell and Sehti is provocative and contains an interesting discussion of record linkage technical issues and available software.

15.4.   Provider Access to Immunization Register Securely (PAiRS) System

PaiRS is a demonstration project implemented by Initiate Systems on behalf of North Carolina's Department of Health and Human Services and the members of the North Carolina Information and Communications Alliance, Inc. (NCHICA), a non-profit organization. The PaiRS database contains records on more than 2 million children aged 0–18 years residing in the State of North Carolina. The database has information on more than 20 million immunizations and is updated once every 2 months. The database was originally created by merging three separate databases: (1) the North Carolina Immunization Registry, (2) the immunization system of Blue Cross and Blue Shield of North Carolina, and (3) the immunization system of Kaiser Permanente, a large health maintenance organization.

The ultimate goal of this work is to provide authorized healthcare providers with the ability to use the Internet to obtain complete information on the immunization history of each child residing in the State of North Carolina.

The purpose of the PaiRS demonstration project was to demonstrate the feasibility of constructing a statewide registry that

• merges the centrally managed State registry with registries of private entities
• provides secure access to immunization records by authorized individuals
• has high-quality, reliable immunization data.

Each record in the database includes the name, address, date of birth and gender of the patient as well as the type and dosage amount of the vaccine administered, the date the vaccine was administered, and the name of the physician or health department that administered the vaccine.

PAiRS has enabled North Carolina's Division of Public Health and other NCHICA members to better understand issues related to data quality, data administration, security, and the integration of data from public and private sources.

15.5.   Studies Required by the Intermodal Surface Transportation Efficiency Act of 1991

15.5.1.   Background and Purpose

In 1991, the Congress of the United States passed the Intermodal Surface Transportation Efficiency Act. This required the US National Highway Traffic Safety Administration (NHTSA) to conduct studies of the effectiveness of (1) automobile safety belts and (2) motorcycle helmets in preventing death and reducing injury severity and property damage arising from motor vehicle accidents.

Police are required to file accident reports on motor vehicle accidents that occur on public roads whenever (1) the estimated amount of property damage meets or exceeds a pre-specified threshold limit or (2) there is at least one serious personal injury. Unfortunately, the police occasionally fail to report all injuries. In addition, the police reports typically lack the exact amount of (1) property damage and (2) the medical costs of treating the injured.

15.5.2.   Study Design

To overcome these problems, NHTSA, as reported in Utter [1995], linked the accident reports of seven states – those "with the most complete statewide crash and injury data" – with medical records from hospitals, emergency rooms, and/or emergency medical services (EMS). The seven States were Hawaii, Maine, Missouri, New York, Pennsylvania, Utah, and Wisconsin. For each, the most recent year of data at the time of the study was selected from 1990 to 1992. This resulted in data on 879,670 drivers of passenger cars, light trucks and vans for the safety belt study; and 10,353 riders of motorcycles for the motorcycle helmet study.

15.5.3.   Results

The safety belt study estimated that safety belts were effective in (1) preventing death 60% of the time and (2) preventing injury 20–45% of the time. NHTSA estimated that the average hospital in-patient charge for an unbelted driver admitted to a hospital was $13,937 compared to $9,004 for a driver who had used a safety belt. Moreover, NHTSA estimated that if all drivers involved in police-reported motor vehicle accidents had been using a safety belt, the seven States would have reaped annual savings of $47 million.

The motorcycle helmet study estimated that helmets reduced the incidence of mortality by 35% and prevented personal injury 9% of the time. When NHTSA restricted the analysis to brain injuries (the injury that the helmet is designed specifically to prevent), however, NHTSA estimated that (1) motorcycle helmets spared 67% of motorcyclists the trauma of a brain injury and (2) the avoidance of each such injury saved about $15,000 in inpatient hospital charges.

15.6. Crash Outcome Data Evaluation System[1]

NHTSA is currently working with a number of states to develop a Crash Outcome Data Evaluation System (CODES). In this work, data on individuals are extracted from police accident reports created at the scene of an accident and then linked to records on (1) statewide emergency medical services and (2) hospital inpatient visits. Other databases are used in the linkage process if available and deemed useful to the project. Such databases include those that contain the following types of information: vehicle registration, driver license, census, roadway/infrastructure, emergency department, nursing home, death certificate, trauma/spinal registry, insurance record, and provider specific data.

The purpose of the study described by Johnson [1997] is to provide more accurate reporting of the severity of injury resulting from motor vehicle accidents in order to (1) reduce the severity of injury, (2) improve medical care, (3) reduce the cost of medical care, and (4) improve highway safety.

Each of the seven States (Hawaii, Maine, Missouri, New York, Pennsylvania, Utah, and Wisconsin) in this study uses distinct data fields to block and link their files. The choices of the fields to be used for blocking depend on (1) both the quality and the availability of the data within the State, (2) the linkage phase, and (3) the databases being linked. Most States used accident location, date, time, provider service area, and hospital destination to discriminate among the events. When the name or an alternative unique identifier was not available to identify an individual, the fields age, date of birth, gender, and description of the injury were the ones used most frequently to identify those involved in the accidents.

The seven States participating in this phase of the study used different schemes to collect data. The quality (e.g., the completeness) of the data varied among these seven States. Nevertheless, after the records were linked, the data were

---

[1] This section is based heavily on Sandra Johnson's "Technical Issues Related to the Probabilistic Linkage of Population-Based Crash and Injury Data."

deemed to be of sufficient quality so that they could be merged to produce (estimated) consolidated effectiveness rates. The States were also able to

- Identify populations at risk
- Identify factors that increased the chance of serious injury
- Identify factors that increased the cost of medical care
- Determine the quality of their State's data.

For each of the seven States, about 10% of the records on individuals constructed from the police accident reports were successfully linked to an EMS report and slightly less than 1.8% of the records on individuals were successfully linked to a hospital inpatient record. This is consistent with the nature of the injuries typically encountered in such accidents. The linkage rates also varied by the severity of the injury, with those more seriously injured more likely to be linked. Because insurance records were more complete/accessible in both New York and Hawaii, about two-thirds of those classified as having "possible injuries" were linked in those two States versus about only one-third in the other five States. Similarly, because hospital data on outpatient treatment were more complete/accessible in New York and Utah, those involved in accidents and classified as having "no injuries" in those two States were much more likely to be linked than in the other five States. Because motorcycle riders typically incur more severe injuries, motorcycle riders were more successfully linked than occupants of cars, light trucks, or vans. Specifically, over 45% of the person-specific accident reports involving motorcycles were linked to at least one injury record in each of Hawaii, Maine, Missouri, New York, and Utah.

16
Constructing List Frames and Administrative Lists

Construction of a frame or administrative list of businesses or agricultural entities for an entire country or a large region of a country involves at least the following major steps. This discussion is based on Winkler [1998].)

1. *Identify existing lists that can be used to create the frame or administrative list –the main list.* It is better to concentrate on no more than 10 lists, choosing those with the most comprehensive coverage and the fewest errors. If an update list does not add enough new entries to the main list, then do not use that update list. If it is likely that an update list will cause too many duplicates to be added to the main list, then do not use that update list. If an update list is on an incompatible computer system or is in an incompatible format (e.g., microfiche), then do not use that update list.
2. *Obtain an annotated layout for each list.* The layout should identify all of the possible values or codes used for each of the fields of the records on the list. The layout should describe how name and address standardization were performed. If no annotated layout is available for a list, then do not use the list.
3. *Make sure that the entries on each update list are in a format compatible with your duplicate-detection and standardization software.* If an update list can't be run through your duplicate-detection or standardization software, then do not use the update list. If a large portion of the entries on an update list can't be standardized, then do not use the list or just use the records that can be standardized.
4. *If resources permit, append each update list to the main list sequentially.* Appending lists one-at-a-time and then completing a clerical review before adding another list usually enables clerical staff to identify and eliminate more duplicate records. Of course, such a process is usually more expensive because it uses more staff resources.
5. *Design the main list so that it has fields that identify the source (i.e., the update) list of each record, the date the source list was obtained, and the date each record was last updated.* Such information is often crucial to eliminating duplicates and to ensuring that current information is on the main list.

6. *Frequently – at least several times per year – check the main list to identify duplicate records and to correct data entries having typographical or other errors.* Because small businesses in the United States (1) go in and out of business frequently, (2) move (thereby changing their address) or (3) change names, it is difficult to keep information on business lists up-to-date. For this reason, it is often helpful with business lists to include a field that denotes the operating status ("open" or "out of business") and another field that indicates whether the entity is a corporate "subsidiary" or the "parent" company. Moreover, if the entity is a subsidiary, then it is important to know the name of the parent company. Finally, it is important to know if the subsidiary reports for itself or if the parent company is the only reporting entity. Frequently, when company "A" buys company "B", company "B" will continue to report for itself for 2 to 3 years after the purchase.

16.1.   National Address Register of Residences in Canada[1]

An address register is a list of residential addresses. Statistics Canada has done lots of work on address registers over the years. The goal of those creating the 1991 Canadian address register was to improve the coverage of the 1991 Census of Canada – that is, to improve the proportion of residences enumerated in the 1991 Census. Specifically, Statistics Canada estimated that the use of the 1991 address register would add 34,000 occupied dwellings and 68,000 persons in the areas for which it would be constructed.

For British Columbia, the address register for the 1991 Canadian Census included all urban population centers as well as some rural areas – approximately 88% of the population. For the other Canadian provinces, the 1991 register was limited to medium and large urban centers. British Columbia was singled out because it had a high rate of undercoverage, 4.49%, in the 1986 Census of Canada.

16.1.1.   Sources of Addresses

Statistics Canada decided to use, wherever possible, four types of administrative databases to obtain addresses for the 1991 address register:

- Telephone billing databases
- Municipal assessment rolls
- Hydro company billing databases
- The T1 Personal Income Tax File.

---

[1] This section is based on Swain et al. [1992].

Unfortunately, all four types of databases could be used only in the provinces of Nova Scotia and New Brunswick and eight major urban centers of Ontario – Ottawa, Toronto, Brampton, Etobicoke, London, Mississauga, Hamilton, and Windsor. Only three types of databases (telephone, hydro, and tax) could be used in the provinces of Newfoundland, Quebec, Manitoba, and Alberta; while only telephone, assessment, and tax databases were used for Regina and the remainder of Ontario. Only telephone and tax databases could be employed in Saskatoon. Finally, British Columbia relied primarily on telephone and hydro databases, but employed motor vehicle, cable, and election databases as well.

16.1.2. *Methodological Approach*

The first step in the process was to standardize and parse the free-formatted addresses from the various source databases. The second step was to create an Address Search Key for each address. The Address Search Key "is an ordered concatenation of all the components of an address and is used" to help identify duplicate records. The third step was to verify or correct, as appropriate, the postal code of each address. The Statistics Canada software (known as the Automated Postal Coding System or PCODE) employed for this purpose verified 78% of the postal codes and corrected 6% of the others. Only 0.0003% of the addresses obtained from the administrative databases lacked a postal code. Postal Codes were a critical input to the next step in the process.

In the next step of the process, Statistics Canada partitioned Canada into 105 geographic areas or "worksites" each of which contained roughly 100,000–150,000 dwelling units based on the 1986 Canadian Census. For medium-sized cities, worksites were created from a single Area Master File (AMF)[2]. For small towns and townships, Statistics Canada created worksites by aggregating geographically adjacent jurisdictions. Finally, for large cities, worksites consisted of part of the large AMF for that city. These 105 worksites contained 20.5 million addresses taken from the various administrative databases, with 22.9 million such addresses outside the AMF areas (i.e., in smaller cities and rural areas) being deleted.

In order to delete addresses included more than once on the remaining portion of the source databases, Statistics Canada used two distinct methodologies. First, they used an exact match procedure to identify and then delete duplicate addresses. This reduced the number of addresses from 20.5 million to 10.1 million. They next used a probabilistic matching scheme via software known as CANLINK. This further reduced the number of addresses to 6.7 million.

---

[2] An Area Master File is a digitized feature network (covering streets, railroads, rivers, etc.) for medium and large urban areas, generally with populations of 50,000 or more. Of interest in the formation of the address register were the street features that contained street name and civic number ranges that could be used to locate individual addresses onto a blockface, the primary linkage. (See p. 130 of Swain et. al. [1992].)

In Step 7, where possible, each address was assigned/linked to a unique blockface[3] using the AMF.

In Step 8, some of the addresses that could not be linked to a unique blockface, in Step 7, were linked to a Census Enumeration Area using Statistics Canada's Postal Code Conversion File – a file that links each Canadian postal code to one or more Census Enumeration Areas. The idea was that this would facilitate manual efforts to assign such addresses to a blockface in Step 10.

In order to facilitate queries, the addresses were loaded in Step 9 onto an ORACLE database management system – a relational database. This consisted of four separate component files: one each for municipalities, blockfaces, streets, and addresses.

The software used in Step 7 was unable to assign (i.e., link) all of the addresses to a blockface. All of these unassigned addresses (i.e., those either the software of Step 8 assigned to a Census Enumeration Area (EA) or those not yet assigned to either a blockface or an EA) were examined manually in Step 10 for possible assignment to a blockface.

Step 11 was a compress step that was applied to all records assigned to a blockface. For each unique value of street name/street designator/street direction within a worksite, all of the corresponding address records were checked for uniqueness using the civic number/apartment number as the key. Where multiple records occurred, all of their pertinent data were consolidated onto a single record, thereby eliminating more duplicate records.

In Step 12, Statistics Canada examined those residual addresses that could not be linked to a unique EA but had been assigned/linked to multiple EAs in Step 8. A complete set of maps was produced by Statistics Canada's Computer Assisted Mapping System (CAM). The clerical staff examined these maps in order to assign the remaining addresses to the appropriate EA wherever possible.

CAM assigns blockfaces to blocks and blocks to Census Enumeration Areas. CAM was used in Step 13 to sort the address register in the order needed for publication. The published address register was a separate booklet for each of 22,756 Census Enumeration Areas and contained a total of 6.6 million addresses.

16.1.3.   Future Goals

Statistics Canada has as a current goal that its national address register will cover 75–80% of Canadian households for the 2006 Census.

16.2.   USDA List Frame of Farms in the United States

The National Agriculture Statistics Service (NASS) of the US Department of Agriculture (USDA) gathers and analyzes information on US crop acreage, livestock, grain production and stocks, costs of production, farm expenditures,

---

[3] By a *blockface* we mean the properties abutting on one side of a street and lying between the two nearest intersections.

and other agricultural items. It collects this information through a number of sampling surveys that it conducts. For example, farm operators and agribusinesses are regularly surveyed in order to obtain statistical estimates of various annual crop yields in the United States.

In its surveys, NASS uses a multiple frame sampling design. Samples are drawn from two different types of sampling frames. One type is a list that consists of the names, addresses, and control data of all known farm operators and agribusinesses. NASS maintains such a list for each of the 50 States of the United States. The control data allow NASS to draw stratified samples, thereby improving the efficiency of its sampling schemes. In the language of survey practitioners, such lists are known as *list sampling frames* or simply as *list frames*.

Because list frames are usually not complete, NASS supplements its list frame information with an alternative type of sampling frame known as an *area-sampling frame*. The area-sampling frame consists of the total land area of the United States partitioned into sampling units consisting of geographical areas defined by easily identifiable boundaries. The area frame provides complete coverage of the geographical area of interest, irrespective of any operational changes. The area-sampling frame enables the analyst to estimate the incompleteness of the list frame.

It is crucial to NASS that no unit in its population of farms be listed more than once on its list frames. If undetected duplicates are present on the list frames, then the probabilities of selection for each unit in the population will be incorrect and the estimates computed using these probabilities may be biased. For these reasons, NASS devotes a considerable amount of time and resources to the maintenance of its list frames.

NASS constructs its list frames from source lists of agricultural operators and operations. These source lists have various origins. Among the most important are the Farm Service Agency (FSA) lists of operators who have signed up for agricultural support programs, lists of members of agricultural producers' associations, lists from State Departments of Agriculture, and lists from veterinarian associations. Operators are often listed on more than one of these lists, a common source of duplication on these list frames. NASS also updates the list frames by the addition of individual records representing operators or operations and by the modification of these records. These updates can also result in the creation of two or more records on the list frames representing the same unit in the population. NASS uses record linkage to detect these duplicates.

To detect duplication created by the modification of records already on a list frame, NASS uses record linkage to compare the list frame to itself – an example of unduplication. This is done annually using naïve software that matches records on Social Security Number (SSN), Employer Identification Number (EIN), or telephone number before sampling from the frame. States typically request a complete unduplication using the current record linkage system every few years or so.

Before NASS adds a new list source to its list frame, NASS parses and standardizes the names and addresses of the items on the new list source. This enables NASS to efficiently use record linkage techniques to compare the items on the list source to those on its current list frame in order to identify items that are on both lists. The operator or operation names not already on the current list frame are then added to thelist frame.

In 1992, NASS decided to replace its list frame maintenance system (known as RECLSS) that was then resident on a mainframe computer with a Unix-based client–server system running a Sybase database. NASS decided to purchase a commercial record linkage software package that would run under Unix and Sybase and have the following capabilities as well:

- It would allow multiple blocking factors and multiple passes of the data.
- It would allow comparison of alphabetic strings using NYSIIS and sophisticated string comparator metrics.
- It would be "statistically justifiable."
- It would parse free-formatted name and address fields.
- It would standardize frequently used terms in names and addresses.

NASS (see Day [1995 and 1996]) did a meticulous job of researching its options and after extensive analysis NASS purchased the AUTOMATCH software developed by Matthew Jaro who was then with a company known as MatchWare Technologies. (MatchWare Technologies was purchased by a company called Ascential Software that was in turn bought out by IBM.) AUTOMATCH includes a component system known as AUTOSTAN that is a flexible name and address standardization system. AUTOMATCH uses the Fellegi–Sunter record linkage methodology, an approach that NASS felt was "statistically justifiable." AUTOMATCH meets all of the other criteria listed above as well.

In order to compare the operation and performance of AUTOMATCH and RECLSS, Broadbent [1996] of NASS tested both systems by linking a list of potential new operations, obtained from the Ohio Producer Livestock Marketing Association (PLMA) with the existing operation on the Ohio list frame. Counts were then made of the number of correct and incorrect decisions made by each system. In the aggregate, AUTOMATCH made 24,207 correct decisions, while RECLSS made 22,455 correct decisions. AUTOMATCH made 61 false matches compared to 855 for RECLSS. AUTOMATCH made 575 false non-matches versus 1,533 for RECLSS. As a consequence, Broadbent [1996] reported "that AUTOMATCH does indeed perform substantially better than the RECLSS system as it is currently used." Finally, Day [1995] concluded that although AUTOMATCH was "the most appropriate choice as the core component for NASS's next record linkage" system, "[t]his recommendation is based on NASS's existing and planned applications and is not in any manner a general recommendation of record linkage software outside of NASS."

16.3.   List Frame Development for the US Census of Agriculture[4]

In the previous section, we discussed some work being done at the USDA. Here, we discuss some parallel work carried out at the US Bureau of the Census for the 1987, 1992, and 1997 US Census of Agriculture. This case study illustrates how sophisticated computer matching techniques can reduce the cost of a sample survey while simultaneously improving its quality.

In 1987 and 1992, the Bureau of the Census wanted to create a list frame (i.e., a list of addresses) of farms in the United States for its US Census of Agriculture. The Bureau of the Census constructed an initial list frame by merging 12 separate source lists of farms – agricultural operators and agribusinesses. These source lists include the Farm Service Agency (FSA) lists of operators who have signed up for agricultural support programs, lists of members of agricultural producers' associations, lists from State Departments of Agriculture, and lists from veterinarian associations. The goal here was to produce a list of addresses by removing duplicates so that each farm appeared exactly once on the list. In 1992, the initial list consisted of approximately 6 million addresses. To facilitate comparison in this discussion, we have multiplied the 1987 proportions by the base of 6 million taken from the 1992 work.

For the 1982 Census of Agriculture as well as for prior censuses, the Bureau of the Census reviewed such lists of addresses manually and made no attempt to even estimate the number of duplicate addresses remaining on its list frame.

For its 1987 Census of Agriculture, the Bureau of the Census implemented "ad hoc" algorithms for parsing names and addresses of farms. For pairs of records agreeing on US Postal ZIP Code, the software used a combination of (1) surname information, (2) the first character of the first name, and (3) numeric address information to identify "duplicates" and "possible duplicates." Of these pairs of records, (1) 6.6% (396,000) were identified as "duplicates" and (2) an additional 28.9% (1,734,000) were designated as "possible duplicates." The "possible duplicates" were then reviewed manually. This clerical effort, encompassing about 14,000 hours of clerical staff time over a 3-month period of time, identified an additional 450,000 duplicate records. The Bureau of the Census estimated that about 10% of the records on the final list frame were duplicates. Because there were so many duplicates, some of the estimates calculated from this survey may be in error.

In 1992, the Bureau of the Census implemented algorithms based on the Fellegi–Sunter model, augmented by effective algorithms for dealing with typographical errors. The resulting software identified 12.8% of the file (about 768,000 records) of the 6 million records as duplicates and an additional 19.7% as requiring clerical review. This time the number of clerical staff hours was reduced by about half, to 6,500 hours over 22 days, and an additional 486,000 duplicates

---

[4] The material in this section is based on Winkler [1995].

were identified. Moreover, this time the Bureau of the Census estimated that only about 2% of the records on the final list frame were duplicates.

Even without the clerical review, the improved 1992 software identified almost as many duplicates as the combined 1987 effort involving both computers and clerical staff. The cost to write the improved software and to do related developmental work in 1992 was only $150,000 – a very good investment, indeed. Nevertheless, there was no actual cost savings in 1992, because the software development cost was equal to the savings from the reduction in the clerical effort. The Census Bureau updated and improved this record linkage software for the 1997 Census of Agriculture at a cost of only $50,000.

16.4.   Post-enumeration Studies of US Decennial Census

The US Bureau of the Census uses record linkage techniques to evaluate the coverage of decennial censuses. Specifically, it estimates the quality of its enumeration by matching individuals counted during the census to those counted in an independent survey conducted after the census. Such a survey is called a *post-enumeration survey*. Another potentially important use of matching in conjunction with post-enumeration surveys is to support a statistical adjustment to Census estimates in order to account for those the Census Bureau fails to count. This is the so-called "Census undercount" that has been the subject of much partisan political discussion in Washington, DC.

The Record Linkage Staff of the Statistical Research Division of the US Census Bureau was created during the 1980s to construct a coherent record-linkage system. The system was intended to be based on a more theoretically sound foundation than previous ad hoc systems. Manual searching for matches is slow, error-prone, and expensive because it entails many clerical staff hours. The technical success of any adjustment procedure depends critically on the ability to match a large number of records quickly, economically, and accurately. Even a few errors may be of critical importance, because population adjustments can sometimes be less than 1%. The goal of this study was to show that the computer could successfully identify the matches that relatively unsophisticated personnel had done in the past, thereby reducing the number of individuals that had to be matched manually during the Census Bureau's future evaluations of the coverage of decennial censuses.

16.4.1.   1985 Test Census in Tampa, Florida

Jaro [1989] describes a test census and an independent post-enumeration survey that the Census Bureau conducted in Tampa, Florida during 1985, while Jaro was with the Census Bureau's Record Linkage Staff. The goal of this work was to test the effectiveness of the new record-linkage system in using the computer to match individuals who would respond in the future to both the decennial census and the post-enumeration survey.

16.4.2. File Preparation

Before attempting to match records, it is usually a good idea to parse and standardize as much free-form information as possible. So, in this work on the Tampa census, Census Bureau staff parsed the names of individuals into distinct fields (e.g., given name, middle initial, and surname). They also parsed the components of the street address into distinct fields and standardized the spelling of common words/abbreviations such as "street" versus "st" and "blvd" versus "bd." Finally, they removed all punctuation symbols from these fields.

16.4.3. Matching Algorithm

Recall that the goal of this matching project was to identify individuals who responded in both the test census and the follow-up (or post-enumeration) survey. So, each record in the file resulting from the test census should only be assigned to one record in the follow-up survey, and vice versa. It turned out that this part of the problem reduced to a problem whose solution is well known to operations research analysts. This problem is a degenerate form of the transportation problem known as the linear sum assignment problem, and is a type of linear programming problem.

16.4.4. Results for 1985 Test Census

In all, 5,343 individuals were matched in the entire process. Of these, the computer identified 4,587 pairs of records as matches , and another 590 were matched via a cursory clerical review. The other 166 matches (i.e., 3.11% of the matches) required extensive clerical review to resolve. The false match rate of 0.174% – just 8 of the 4,587 computer-identified matches were false matches – was extremely low. In sum, this project was a major success in that (1) only 3.11% of the records that were matched required extensive clerical review and (2) the false match rate was only 0.174%.

16.4.5. Results for 1990 Census

This record linkage system was subsequently applied to both the 1988 dress rehearsal for the 1990 Census and the 1990 actual Census itself. The results are summarized in the table below:

Resources required for post-enumeration matching work of US decennial census

Census	1980	1988 dress rehearsal for 1990 census	1990 census
Number of clerks	3,000	600	200
Number of months required to complete work	6	1.5	1.5

(Continued)

(Continued)

Census	1980	1988 dress rehearsal for 1990 census	1990 census
False match rate	5.0%	0.5%	0.2%
Percentage of matches identified by the computer	0%	70%	75%

The cost to develop this record linkage software was about $3 million. This was a terrific investment, particularly because this software has been used for a number of subsequent projects. This success is especially impressive because so many other large-scale government software development projects have ended in total failure.

17
Social Security and Related Topics

17.1. Hidden Multiple Issuance of Social Security Numbers

17.1.1. Introduction

In the mid-1990's, Bertram Kestenbaum [1996] of the Social Security Administration (SSA) conducted a record linkage study whose principal goal was to identify persons who were born in the United States during 1919 and were subsequently issued more than one Social Security number. In addition to improving the quality of one or more of SSA's databases, the results of such a study can be used to improve SSA's estimate of the extent to which people receive program benefits at least equal to the taxes they've paid during their work-lives – the "money's-worth" issue.

17.1.2. Background on SSA Databases

We begin by describing two major SSA administrative files that were important to Kestenbaum's study. The first "is the NUMIDENT file of initial applications for a social security number and card, reapplications to replace lost cards or to correct identifying information (such as changes in surname upon marriage), and certain records of claims to benefits. The NUMIDENT file is also used to house death information which became available to the Social Security Administration." Each "NUMIDENT record has extensive personal information: name, name at birth, date of birth, place of birth, sex, race, parents' names, and date of death, if known deceased."

The second "major administrative file is the Master Beneficiary Record, the file to entitlements to retirement, survivorship, and disability program benefits." "The personal information in this file, which includes name, date of birth, sex, race, and date of death, if deceased, is not quite as extensive as in the NUMIDENT."

17.1.3. The Data

Kestenbaum chose to work with a 1-in-100 sample of those born in 1919 and who were in the NUMIDENT file because they had applied for a Social Security number. The selection was based on patterns of digits in the Social Security Number. Under this sampling scheme, the *undetected* issuance of multiple Social Security numbers posed a twofold problem. First, Kestenbaum needed to identify individuals in his 1-in-100 sample whose main accounts[1] were *outside* his 1-in-100 sample. Second, Kestenbaum needed to identify individuals in his 1-in-100 sample whose main account was *inside* his 1-in-100 sample but who had at least one additional account as well. In both situations, this required transferring the information from the additional account(s) to the main record.

Kestenbaum used three deterministic (i.e., exact) matching schemes and one probabilistic matching scheme to identify duplicate records.

17.1.4.   Identifying Duplicate Records Using Deterministic/Exact Matching Schemes

Kestenbaum's first initiative was to exact-match (1) records of those in the 1-in-100 sample who had not yet applied for benefits with (2) a file of 2 million records from SSA's Master Beneficiary Record (MBR) database of those not in the 1-in-100 sample; these 2 million records represent individuals born in 1919 who had already applied for SSA retirement, survivorship, or disability benefits. Kestenbaum used various combinations of identifying fields (e.g., Social Security number, name, and date of birth) to perform this exact match. The 144 records (representing an estimated 14,400 ($= 144 \times 100$) records in the original database) which successfully matched were dropped from the 1-in-100 sample.

His next initiative was to examine the records on "suspect" number holders – that is, those who had not applied for benefits through 1992 (73 years after his/her birth), were not working, and were not identified as "deceased" within any SSA database. Kestenbaum then used the NUMIDENT query system to which he manually entered a name and date of birth (with certain tolerances) and received back a printout of matching NUMIDENT records. Using this manual facility, Kestenbaum discovered (1) 325 multiple number holders whose record could be deleted from the 1-in-100 sample because a newly discovered number outside the 1-in-100 sample was the main number and (2) 45 multiple number holders whose newly discovered number was less active, but whose activity under this account needed to be added to their account within the 1-in-100 sample.

---

[1] For individuals issued more than one Social Security number, Kestenbaum chose to identify one as the "main" number. For example, if an individual received benefits under a single number, then Kestenbaum chose that number as the "main" number. If the individual had not yet received any benefits, but had paid taxes for 30 years under one number and three years under a second number, then Kestenbaum designated the first number as the "main" number.

Next Kestenbaum obtained a computer file consisting of 6 million NUMIDENT records with year of birth of 1919. He sorted these records according to the set of values for certain combinations of identifiers. He did an exact match that matched records in the 1-in-100 sample to those in the rest of the file. Here he matched *exactly* on the combination of identifiers. This process produced 350 duplicate records, resulting in (1) the deletion of 127 records from the 1-in-100 sample and (2) the addition of activity to 223 records within the 1-in-100 sample.

17.1.5. *Identifying Duplicate Records Using Probabilistic Matching Schemes*

Having just completed a discussion of the exact matching schemes that Kestenbaum used to identify multiple Social Security numbers issued to those born in 1919, we next describe his efforts to employ probabilistic matching techniques to identify such numbers. Specifically, we describe his use of probabilistic matching techniques on the combined NUMIDENT and Master Beneficiary Record files. His basic approach was a variant of the frequency-based record linkage (matching) scheme we discussed in Section 9.2.

Kestenbaum [1996] began by noting that the technique of probabilistic matching is especially valuable when (1) the identifiers used in the linkage are prone to change, as is the case for surnames of females, and (2) the identifiers are frequently missing or in error. The NUMIDENT, in particular, has serious data quality shortcomings, largely for two reasons. First, the creation of the NUMIDENT in the mid-1970s was an enormous data entry task lasting several years. The information was manually entered from hundreds of millions of paper forms. The size of the operation made conditions less than conducive to quality control. Second, it was the practice in the old paper environment, that in the adjudication of claims for benefits, the account number application form would be physically removed from the file to become part of the packet of documents provided to the adjudicator. The account number application form would then be replaced in the file by a claims form that generally lacked much of the information on the original form. Often even the sex code was missing, forcing SSA to depart from its standard practice of partitioning the linkage operation into two: one for males, the other for females. Separate-sex treatment reduces the size of the linkage, as well as allows for disagreement on surname to be of less importance for females than for males, as it should be.

For his application, Kestenbaum [1996] used the following identifiers: month of birth, day of birth, surname (including surname at birth), given name, middle name, sex, race, first five digits of Social Security number (which reflect geographic and temporal information as of the time of issue), State and place of birth, vital status, date of death, mother's maiden name, mother's given name, mother's middle initial, father's given name, and father's middle initial. For each Social Security number, Kestenbaum [1996] produced a record for every unique combination of these identifiers; beginning with approximately 6 million NUMIDENT records and approximately 2 million Master Beneficiary Records for persons born in 1919, close to 30.75 million unique records were produced for linkage.

In the preprocessing stage, Kestenbaum expended considerable effort on standardizing place names, at least the more common ones. For example, before standardization, Brooklyn, NY could appear as BROOKLYN, BKLYN, or KINGS (COUNTY). Another interesting preprocessing initiative was his attempt to identify which female middle names were actually maiden names, by comparing them to a list of common female given names.

Records with social security numbers in the 1-in-100 sample constituted "File A"; all others became "File B." File B contained about 30.5 million records; the remaining 250,000 records were on File A. This was indeed a large application of probabilistic linkage.

As his reference file for calculating the frequencies of comparison outcomes among true matches, Kestenbaum used record pairs that had previously been identified in SSA files as instances of multiple issue. Using random numbers, Kestenbaum fashioned a very large file of randomly linked record pairs to estimate the frequencies of comparison outcomes among random pairs of records.

Comparisons were made first on a global basis, and then – for pairs scoring above a chosen threshold on the global basis – on the more discriminating value-specific basis, as well. Partial agreements were recognized to a very large extent. The linkage was repeated under eight blocking strategies.

To make the transition from global to specific, lookup tables were constructed containing the frequencies of identifier values in File B. In order to speed up the data processing, Kestenbaum deleted from the lookup table all values that did not appear in File A. With respect to given names, for example, Kestenbaum was able to reduce the size of the lookup table from about 135,000 records to less than 5,000.

These probabilistic linkage schemes enabled Kestenbaum (1) to delete another 227 records from his 1-in-100 sample because they were on individuals whose main number was outside the 1-in-100 sample and (2) to identify 371 records within the sample that needed information added from duplicate records outside the sample.

17.1.6. Conclusion

In his conclusion, Kestenbaum said he was pleased that probability matching was able to identify duplicate records he had not been able to find using other schemes. He also noted that his "modest success from probabilistic linkage would have been more impressive had [he] not uncovered several hundred" duplicate record matches earlier using exact matching schemes.

17.2. How Social Security Stops Benefit Payments after Death

The Social Security Administration (SSA) has a continuing concern about erroneous payments of benefits to deceased individuals under two of the benefit programs that it administers: Old Age Survivors and Disability Income (OASDI) and Supplemental Security Income (SSI). A few years ago, the issue concerned the act of cashing checks sent to a deceased beneficiary. In many cases, there would be a conscious decision by a relative or other individual to commit a fraudulent act. Now, as the majority of payments are made electronically, more of the burden is on SSA to recover these payments made after the death of a beneficiary, because the decedents' survivors may not even be aware of such direct deposits to the decedents' bank accounts.

Of the approximately 2.3 million people who die in the United States each year, about 2.0 million are SSA beneficiaries. Most deaths are reported to SSA by relatives, friends, or funeral homes. SSA estimates that such reports encompass about 90% (2.07 million) of these deaths. Postal authorities and financial institutions report another 5% (115,000) of such deaths.

SSA relies on a sophisticated computer system to identify the remaining 5% (115,000) of deaths that are not reported. This system – known as the Death Alert, Control, and Update System (DACUS) – relies upon death reports from sources both inside and outside SSA. The external sources include Federal agencies such as the Centers for Medicare and Medicaid Services (CMS) and the Veterans Administration (VA) as well as State agencies such as bureaus of vital statistics (BVS) and social service agencies.

The internal sources include SSA's Beneficiary Annotation and Communication Operation (BACOM). This system cross references an individual's own Social Security number with his/her Social Security claim number in case he/she is entitled to benefits from another individual's work record. Examples of this include surviving spouses, ex-spouses, and children of deceased workers. (Also considered an internal source is any action that updates the MBR and passes through SSA's Daily Update Data Exchange (DUDEX).) DUDEX collects information from daily SSA transactions that need to be transmitted to other Federal agencies such as the US Department of the Treasury, the Railroad Retirement Board, or the Centers for Medicare and Medicaid.

In essence, DACUS compares the information on these death reports to the corresponding information on SSA's payment files – the Master Beneficiary Record (MBR), the Supplemental Security Record (SSR), and the Black Lung Payments System (BLUNG). The Master Beneficiary Record is the master file for participation in the Social Security and Medicare programs. The Supplemental Security Record is the master file for participation in the Supplemental Security Income program for impoverished aged persons. The Black Lung Payments Systems is the master file for participation in the Black Lung program for disabled mine workers. Finally, a fourth Social Security database, NUMIDENT, is the file of the hundreds of millions of applications for an original or replacement Social

Security card. The NUMIDENT also contains corrected identifying information such as changes in surname upon marriage. Finally, the NUMIDENT is the repository (i.e., database) for the death information reported to the agency. This portion of the NUMIDENT is known as the Death Master File (DMF). The NUMIDENT has extensive personal information: name, name at birth, date of birth, place of birth, sex, race, parents' names, and date of death, if known. The DMF contains the following personal information: name, Social Security Number, last known address, dates of birth and death, and the State in which the individual initially applied for a Social Security Number.

When DACUS receives a DTMA (death record to add) or a DTMC (death record to correct), DACUS compares these records to the corresponding record, if any, on each of the following databases – in the order listed below:

- MBR
- SSR
- BLUNG
- NUMIDENT

If the records match exactly on (1) the Social Security number, (2) the first six letters of the last name, (3) the first four letters of the first name, and (4) the month and year of birth, then the records are said to have an *ideal match*. Alternatively, if (1) the records match exactly on the Social Security number, (2) the records match exactly on the first six letters of the (SOUNDEX version) of the last name, (3) the records match on the first initial of the first name, (4) either (a) the month of birth is the same and the years of birth are no more than 3 years apart or (b) the month of birth is different but the reported birth dates are no more than 12 months apart, and (5) the sex code on the MBR, if valid, matches the sex code on the State record, if valid, then the records are said to have an *acceptable match*.

If (1) there is at least an acceptable match, (2) there is no conflicting information on the date of death, and (3) no payments have been made after death, then DACUS records the death information on the DMF portion of NUMIDENT. However, if the comparison indicates that (1) at least one payment was made after death or (2) there is conflicting information about the date of death, then DACUS asks the appropriate SSA field office to obtain additional information. This is called an "alert." In such a situation, the field office may take action, as appropriate, to terminate future payments, recover erroneous payments made after the death of the beneficiary, and/or correct records in one or more of SSA's computer systems.

At the end of each quarter of the calendar year, SSA creates a supplemental electronic file containing newly created death records and any changes or deletions to the DMF. This supplemental file is made available to interested members of the public, including life insurance and annuity companies that use this file to update their own administrative databases.

Certainly, Social Security would prefer that no payments were made after the deaths of its beneficiaries. Administratively, this is impossible. In fact, each year, hundreds of thousands of payments are made after death. Fortunately, the overwhelming majority of such payments are returned, not negotiated, or recovered as overpayments. As hard as the Social Security staff may try to have a perfect death recording system, they still have to deal with practical realities. One is that even today a not-insignificant portion of death certificates lacks a Social Security Number.

17.2.1. Related Issues

A closely related issue involves the erroneous payment of food stamps to deceased individuals in the United States – a program under the auspices of the US Department of Agriculture (USDA).

In addition, under US law, individuals incarcerated in prisons in the US are not eligible for government benefits under the USDA food stamp program or any of the programs administered by the Social Security Administration (i.e., pension benefits, disability income benefits, or Supplemental Security Income benefits payments).

So, federal government agencies have responsibility in these areas as well.

17.3. CPS–IRS–SSA Exact Match File

The 1973 CPS–IRS–SSA Exact Match Study was carried out jointly by the US Bureau of the Census and the Social Security Administration (SSA). The goals of this study included the following:

1. Improve policy simulation models of the tax-transfer system.
2. Study the effects of alternative ways of pricing social security benefits.
3. Examine age reporting differences among the three sources of data.
4. Summarize lifetime covered earnings patterns of persons contributing to social security.
5. Obtain additional information about non-covered earnings – that is those not subject to Social Security taxes.

The primary data source for the study was the Current Population Survey (CPS). The study attempted to link the survey records of all of the individuals included in the March 1973 CPS to (1) the earnings and benefit information contained in their SSA administrative records[2] and (2) selected items from their 1972 Internal Revenue Service (IRS) individual income tax returns.[3]

---

[2] More precisely, this information was taken from SSA's Summary Earnings Record (SER) and Master Beneficiary Record (MBR) databases.
[3] More precisely, this information was taken from the IRS Individual Master Tax File for 1972.

The CPS is a monthly household survey of the entire civilian non-institutionalized population of the United States – that is the 50 States and the District of Columbia. Armed Forces members are included if they are living (1) on post with their families or (2) off post. The CPS collects demographic data as well as data on work experience, income, and family composition. The March 1973 CPS consisted of about 50,000 households and more than 100,000 individuals aged 14 or older within those households.

In order to match the individuals surveyed in the March 1973 CPS to their corresponding IRS and SSA data, their personal identifying information had to be obtained and perfected. This objective was accomplished by taking SSN's and names, and so on, from the CPS Control Card.

The CPS Control Card, at the time this work was done, was simply a piece of cardboard used by the CPS interviewers. It had been redesigned in October 1962 for matching, but had been used previously by interviewers merely to control their own interviews. (In the CPS, a rotation design is used with eight interviews over a 16-month period. The respondent is part of the survey for four consecutive months, out of the survey for the next 8 months, and then in the survey for the next 4 months.)

For the CPS Match project, all identifying control items were keyed and then validated against the corresponding items in the SSA files. When the SSN's were judged invalid or were missing, an attempt was made to obtain valid SSN's via manual search procedures.

After the SSNs were confirmed or supplied, when possible, they were used to match to the IRS extract that the Census Bureau had for its current population estimates program. SSN matching was also carried out at Social Security to obtain earnings and benefit records.

Despite resource and other operational constraints, the machine-validation and manual-search procedures resulted in about 90,000 potentially usable numbers from the $100,000+$ persons aged 14 or older included in the March 1973 CPS.

The Exact Match Study produced six computerized databases that were documented and distributed to interested members of the public for general use. They are as follows:

1. March 1973 Current Population Survey – Administrative Record Exact Match File

   This file contains demographic data from the March, 1973, CPS as well as extracts from (1) SSA earnings and beneficiary records and (2) IRS tax returns for the calendar year 1972.

2. March 1973 Current Population Survey — Summary Earnings Record Exact Match File

   This file contains basic demographic data from the March 1973 CPS as well as CPS and SSA earnings items.

3. 1964 Current Population Survey – Administrative Record Pilot Link File

This was prepared to give researchers a means of introducing a historical dimension to their analyses of the 1973 Exact Match Study Files. Therefore, to the maximum extent possible, the items on this 1964 file have been made identical to those on the 1973 files.

4. June 1973 Current Population Survey Exact Match File

This file contains demographic data as well as income, food stamp, and other information obtained from those individuals the CPS interviewed in both March and June of 1973. This sample consists of approximately 25% of the individuals interviewed in the March survey.

5. Longitudinal Social Security Earnings Exact Match File

This is a longitudinal extract from Social Security earnings for matched adults in the 1973 study sample.

6. 1972 Augmented Individual Income Tax Model Exact Match File

This is an independent sample containing the public-use data made available by the IRS in its 1972 Statistics of Income Match File. It is augmented by selected demographic data extracted from SSA beneficiary records.

17.4.   Record Linkage and Terrorism

A recent article, Gomatam and Larsen [2004], having the title of this section, asks: "How can databases be linked to determine whether collective behavior suggests threatening activities?" These authors adduce the following example: "if passport applications, plane reservations, applications to carry a firearm, and arrest records can be linked, then it might be possible to identify a person or persons engaging in potentially dangerous planning before a terrorist activity occurs, locate them, and prevent them from executing their plans."

They also state that the US Department of Homeland Security plans to analyze information provided by law enforcement and intelligent agencies including the Central Intelligence Agency (CIA), the Federal Bureau of Investigation (FBI), and the National Security Agency (NSA). They go on to report that "The Information Awareness Office (IAO) of the US Defense Advanced Projects Research Agency (DARPA) plans to connect transactions by quickly searching and correlating entries in legally maintained databases as part of its Terrorism Awareness (TIA) System."

Part 4
Other Topics

18
Confidentiality: Maximizing Access to Micro-data while Protecting Privacy

Access to micro-data sets that come from business and survey/census operations has grown greatly, particularly in the last decade or so. The access to such datasets presents a new set of issues. Such issues are the focus of this chapter. Until now we have been concerned with optimizing data quality, subject to constraints like time or money. To these constraints we need to add another – the need to protect the "privacy" of the individuals or entities whose data are being used. We put privacy in quotes because in common usage the obligation applies only to individuals. However, we also need to include other legal entities, like businesses. When data are shared to complete a business transaction, like the use by a customer of his/her credit card to purchase something, there is an implicit or increasingly explicit commitment by the receiving entity to keep that information "in confidence."

In this chapter, we discuss the data quality implications of keeping a promise of confidentiality, however made. Most of the chapter is concerned with the idea of public-use files, developed for statistical purposes by Federal agencies. However, privacy issues are important in the private sector as well. One key area concerns individual medical records. This is the subject of Section 18.5.

Privacy issues may also be important when dealing with data collected by radio frequency identification devices (RFIDs[1]). There are two reasons that privacy might be an issue with RFID data. First, it is now relatively easy to use RFIDs to automatically collect large quantities of data in real time. Second, it is possible to use record linkage techniques to merge data collected by different entities at different times and at different locations.

Every federal statistical agency is tasked with collecting high quality data to facilitate informed public policy and enable statistical research. But every federal agency is also legally obligated to protect the confidentiality of the people or other entities about whom/which they have collected the data – be they persons, organizations, or businesses. These objectives inherently conflict, however, because any technique designed to protect confidentially must, by definition, omit or change

---

[1] The purpose of an RFID system is to enable data to be transmitted by a mobile transponder to an RFID reader. The data are then processed according to the needs of the application. The transponders or tags are typically attached to or incorporated into a product, animal, or person for the purpose of identification. RFIDs are found on tags used to pay tolls automatically at toll booths as well as in library books.

information that could lead to identification. This conflict is becoming ever more complex as both technological advances and public perceptions change in the 21st century information age. Because the omissions and/or changes usually degrade the data, the challenge is to nevertheless keep the data fit for use.

What agencies typically disseminate first are tabulated (i.e., summary) data. But the analyses that can be done with such data are limited. The types of rich statistical analysis that shed light on complex but important questions often need micro-data because such questions cannot be answered via tabulated data alone. To meet this crucial societal need, statistical agencies produce public-use micro-data files.

To develop public-use micro-data files, Y, agencies use various methods to change or distort the values in the original micro-data file, X, in order to make individual information much more difficult to re-identify. These procedures are often called *masking* procedures. Because every federal agency is also charged with protecting the confidentiality of its survey respondents – important not only for ethical reasons but also to preserve public willingness to continue to supply data – these agencies need to ensure that public users cannot associate a masked record in file Y with an original record in file X (i.e., re-identify individual information).

We begin this chapter with several issues that apply whatever the masking techniques used. Section 18.1 emphasizes the need to ensure data quality in the original file before developing the public-use version. Section 18.2 discusses the important role of file documentation in helping to choose a suitable masking method. Section 18.3 makes the point that the agency in question must explore the possibilities of re-identifiability in its public-use files before their release. In Section 18.4, we describe a number of elementary masking techniques that have been considered by various federal statistical agencies, with brief assessments of their strengths and weaknesses. In Section 18.5, we discuss ways to protect the confidentiality of medical records. In Section 18.6, we describe some recently developed masking techniques that are more complex and worth considering for future use.

18.1. Importance of High Quality of Data in the Original File

The quality of the data within the original file, X, has a major impact on the quality of the data within the public-use file, Y. Since the quality of micro-data affects their use, it is in everyone's interest that the original data file is of high enough quality to be reliably used for multiple analyses.

We illustrate this in the following elementary examples.

Example 18.1

In the original file, X, 1% of the records may be on children under the age of 15 who are listed as being married.

Example 18.2

In the original file, **X**, 0.1% of the records may be on female spouses who are 50 or more years older than their husbands.

Example 18.3

In the original file, **X**, six companies in a particular industry may have annual gross revenue above an amount of $K, when it is known from external sources that only two of these companies have such a high annual gross revenue figure.

   Observation: In all three of these examples, the anomalies from the original dataset, **X**, will be preserved in the masked (public-use) dataset, **Y**, unless steps are taken to "correct" such errors in the original micro-data file. Revealing such errors after release of a public-use file will, among other problems, adversely affect the releasing agency's reputation – and quite plausibly its ability to continue – to collect the data necessary for its mission. An ounce of *care* is worth a pound of *cure*.

18.2.   Documenting Public-use Files

To evaluate the quality of a public-use micro-data file, we need to understand how a masking method is intended to (1) preserve the file's analytic strength while (2) reducing identifiability. It is important in addressing point (1) for the agency creating the file to develop detailed written documentation of

• the analytic properties of the public-use micro-data file
• the limitations of the public-use file
• the model used to create the public-use file
• guidelines for analyzing the public-use file.

   If the agency is able to describe the specific analytic properties of the files in this way, it should be able to determine aggregates that are needed in the analysis. If a set of aggregates is known, then it is straightforward to check whether aggregates based on **Y** are approximately equal to aggregates based on **X**. If the file **Y** does not preserve the analytic properties of **X**, the file **Y** should not be made public. If one or two critical analytic properties that are in **X** are not in **Y**, then it may be possible to *refine* the procedures so that a new file **Y1** is produced that satisfies these analytic properties.

18.3.   Checking Re-identifiability

If file **Y** has justifiably valid analytic properties (i.e., objective (1) is attained), then the agency needs to address point (2) by checking re-identifiability. With re-identification, an entity's name or other unique identifying information is associated with one or more records in **Y**. Individuals who might try to re-identify information in this fashion are referred to as *intruders*. There are many of them out there in the ether. The job of the agency charged with developing the public-use data file is to mimic the types of processes the intruders might use, in order to be able to take preventive action before releasing the public-use files.

Example 18.4

In a simple situation, records in a public-use file **Y** may contain variables such as US Postal ZIP code and date of birth. If we have access to an appropriate voter registration list, then we can use the ZIP code and date of birth together with record linkage software to link a specific name and address with some of the records in **Y**. Although certain *quasi-identifiers* such as ZIP and date of birth are no longer in most public-use files, other combinations of variables (fields) can serve as quasi-identifiers.

On a slightly more sophisticated level, we can use record linkage or nearest-neighbor software to match the original file **X** against the public-use file **Y**. This will provide us (as public-data producers) with the worst-case scenario for re-identifying records (or information within records). This worst-case scenario will help identify records/information that we might not have initially thought could be re-identified. We can extrapolate the re-identification rate downward based on our understanding of the original dataset **X**, the analytic and other properties in **Y**, and external data **Y1** (having name and address, etc.) that may be available to an intruder.

One concern with having the micro-data on file **Y** available to the public is that these external data are increasingly becoming of such high quality and unmasked that even *manual re-identification* is often possible. Another is the increasing sophistication of individuals who have (1) knowledge of population data and analytic constraints as well as (2) the ability to develop record-linkage software to facilitate re-identification. Whereas a public-use file may have 10–20 (or more) variables, it may only take 6–8 variables for re-identification. Merely changing or masking a few variables in going from file **X** to file **Y** will not prevent re-identification. A key advantage of using record linkage software in a secure environment for comparing data **X** to data **Y** is that it quickly identifies the subsets of variables that can be used for matching. The subsets vary for each pair of records. It is not unusual for *several* subsets to allow *accurate re-identification* with a pair of records.

If we know the analytic properties of the public-use data **Y** and the characteristics of the original population **X**, then we may be able to re-identify some records in a straightforward equations-unknowns procedure using the analytic constraints as follows:

Step 1: For each analysis, we determine the set of aggregates and other constraints (such as positivity or additivity) that hold for a set of micro-data **Y**.

Step 2: We determine the number of fields, the number of value-states of each field, and the number of records.

Step 3: We use the equations of Step 1 and the unknowns of Step 2 to determine whether there are sufficient degrees of freedom on the masked data **Y** so that some of the data can be re-identified via analytic means.

If the number of equations (constraints) is more than half the number of unknowns, then an intruder may be able to re-identify some records. The re-identification can be due to the fact that certain combinations of values of continuous variables may be in the tails of distributions, or that certain combinations of the variables of the records in **Y** may yield outliers that are very different from other records.

Example 18.5

As an intuitive example, we may want a number of moments to be preserved. If certain outliers from a single variable or a combination of two variables are removed or adjusted substantially, then some of the moments will change substantially. With substantial change in some moments, certain analyses in **Y** will no longer even approximately reproduce analyses in **X**.

Example 18.6 is an example of manual re-identification.

Example 18.6

An intruder may construct a file I_1 from voter registration lists, add in information from county tax records about mortgage amounts, tax payments, and other publicly available information, and also locate other information that can be merged into file I_1. The statistical agency produces a potential public-use file **Y** that breaks out various quantitative data such as income by age, race, sex, and ZIP. After a simple tabulation, the agency realizes that there are two Asian ethnicity women over 70 years of age with high incomes that might be re-identified using a file such as I_1 or just by knowing the characteristics of certain individuals. The agency deals with this by collapsing various age ranges associated with certain races and truncating high incomes within categories so that five or more records in the *coarsened* file **Y1** obtained from **Y** cannot be separated from each other easily.

If we (as public-data producers) are able to re-identify some records using record linkage or analytic constraints or a combination of the two, then we may need to *coarsen* the data **Y** so that fewer records and less information can be re-identified. When coarsening the data **Y** to data **Y1**, the agency must check whether **Y1** still has the requisite analytic properties.

We consider another example.

Example 18.7

We assume that a file **X** consists of 10,000 records on individuals living in a particular State and that each record has seven fields: ZIP Code of Residence, Profession, Education level, Age, Gender, Race, and Income. We further assume that the Profession variable is partitioned into 1000 categories, the Education variable is partitioned into 14 categories, and Income is rounded to thousands of dollars. In this case it is possible that two variables alone may be sufficient to establish a re-identification. For instance, if there is only one surgeon in a particular ZIP code, then it may be possible to associate the name of the surgeon with the income in the file.

Observation: Several straightforward things might be done to overcome this problem. The Profession categories might be partitioned into only 50 (collapsed) categories in which all types of physicians might constitute a single category. In the collapsed data **Y**, there may be several physicians in the ZIP code and re-identification might not be easy. If the doctor has exceptionally high income, then the income might be replaced with a top-coded income of, say, $400,000 making it more difficult to associate an income with a particular doctor. If Education and Race are blanked (given a single neutral value that effectively removes the Education and Race variables from the file), then it may still be possible to re-identify the doctor using ZIP, truncated income, age, and sex. Although it is not possible to find the exact income of the doctor, the fact that we can reasonably identify the doctor's data (in a severely altered form) in the masked file **Y** may still make re-identification an easy task.

From the above example, we can see that masking can place severe limitations on the analytic properties of micro-data. And even when the masking is severe, it may still be possible to re-identify some information in some records.

18.4.   Elementary Masking Methods and Statistical Agencies

Statistical agencies have typically adopted masking methods simply because they are *easy to implement*. But the easiest-to-implement methods have seldom, if ever, been justified in terms of preserving analytic properties or even of preventing re-identification. In extreme situations, such easy-to-implement masking methods may yield a file that cannot be used for analyses at all and yet still allows some re-identification.

In this section we discuss a number of easy-to-implement masking methods that have never been demonstrated to preserve analytic properties.

18.4.1.   Local Suppression

One way of masking records is for the statistical agency to deliberately delete data items from the original database. This causes missing data (a type of item non-response) that makes the masked file unsuitable for use in a statistical or data mining software package. This method is also referred to as *local suppression*.

Example 18.8

We construct an artificial dataset **X** that has properties typical of many real data files. Here, **X** consists of 100 records each of which has four continuous

TABLE 18.1. Sample pairwise correlation coefficients
based on original data

Variables	Variables			
	X_1	X_2	X_3	X_4
X_1	1.000	0.895	0.186	−0.102
X_2	0.895	1.000	0.218	−0.065
X_3	0.186	0.218	1.000	0.329
X_4	−0.102	−0.065	0.329	1.000

TABLE 18.2. Sample pairwise correlation coefficients
for locally suppressed (i.e., deleted) data

Variables	Variables			
	X_1	X_2	X_3	X_4
X_1	1.000	0.875	0.271	−0.125
X_2	0.875	1.000	0.122	0.045
X_3	0.271	0.122	1.000	0.331
X_4	−0.125	0.045	0.331	1.000

variables (fields). The variables X_1 and X_2 are highly correlated and variables X_3 and X_4 are correlated. The variables are distributed in the following ranges: $1 \leq X_1 \leq 100$; $17 \leq X_2 \leq 208$; $1 \leq X_3 \leq 100$; and $61 \leq X_4 \leq 511$. The sample pairwise correlation coefficients of the original data are given in Table 18.1.

We perform local suppression by randomly deleting 20 values of X_1, 26 values of X_2, and 25 values of X_3 in such a manner that no record has more than one value deleted and we do not allow simultaneous blanking; we then compute the sample correlation coefficients for the variables with the deleted entries. The results are summarized in Table 18.2.

Some of the correlation coefficients have changed dramatically from Table 18.1 to Table 18.2; for example, the correlation coefficient of variables X_2 and X_3 has decreased from 0.218 to 0.122 – almost a 50% decrease.

18.4.2. Swapping Values within Specified Individual Fields (or within Several Fields Simultaneously) Across Records

Swapping is a masking technique that is easy to implement. However, even if only a few values of a few fields are swapped, then the resultant masked file has no analytic properties beyond the means of individual fields that are preserved. No correlations, regressions, or loglinear modeling properties are preserved.

TABLE 18.3. Sample pairwise correlation coefficients after swapping

Variables	Variables			
	X_1	X_2	X_3	X_4
X_1	1.000	−0.876	0.936	−0.942
X_2	−0.876	1.000	−0.980	0.935
X_3	0.936	−0.980	1.000	−0.956
X_4	−0.942	0.935	−0.956	1.000

Example 18.9

We begin with the original, but artificial, dataset of the previous example. We swap a number of values across our four fields and compute the sample pairwise correlation coefficients as shown in Table 18.3.

Observation: Correlations are *almost instantly destroyed* with swapping. But the re-identification rate is effectively 0 (calculation not shown).

In some situations (e.g., Kim and Winkler [1995]), swapping has been applied in a very limited manner – only on the most easily re-identified 0.5% of the records in a public-use file Y. Although additional controls were placed on the swapping so that many statistics were preserved (i.e., controlled) within a large set of subdomains, the statistics in the other subdomains were instantly destroyed. For more details on swapping, see Dalenius and Ross [1982] or Fienberg and MacIntyre [2005].

18.4.3. *Rank Swapping Values in Records*

Rank swapping provides a way of using continuous (quantitative) variables to select pairs of records for swapping. Instead of requiring that variables match (i.e., agree exactly), they are defined to be close based on their proximity to each other on a sorted list on the continuous variable. Records that are close in rank on the sorted variable are designated as pairs for swapping. Frequently, in rank swapping the variable used in the sort is the one that is swapped.

Single-variable rank swapping (Moore [1995]) involves sorting the values in individual fields (continuous variables) and swapping values within $p\%$ of the record of interest. To be more specific, we suppose that a dataset has 1,000 records and that $p\%$ is chosen to be 5%. We denote the values of the i-th variable by $(X_{i,1}, X_{i,2}, \ldots, X_{i,1000})$. We sort the values of the i-th variable in ascending order and let $(Y_{i,1}, Y_{i,2}, \ldots, Y_{i,1000})$ denote the result. Then, the only values that can be swapped with $Y_{i,k}$ are those within 5% (i.e., within 50 records in either direction) of $Y_{i,k}$ – in other words, those values between $Y_{i,k-50}$ and $Y_{i,k+50}$. Ad hoc methods must be used to treat the tails of the distributions (either the beginning or the end).

If the swapping percentage p exceeds a small value (say 1 or 2%), then the correlations can be severely affected. Because rank swapping is a special case of the micro-aggregation method (discussed in Section 18.4.6), re-identification is quite straightforward.

As the value of p goes toward 0, the amount of distortion in the swapped (masked) file **Y** decreases.

Example 18.10

In this example, we again begin with the original dataset of Example 18.8. We then perform three types of rank swapping – at 5, 10 and 20%. The results are summarized in Tables 18.4–18.7.

Observation: Rank swapping at 5% preserves correlation coefficients somewhat. But at 10 and 20% rank swapping rates, the deterioration of correlation coefficients is substantial.

In some applications of rank-swapping (e.g., Winkler [2002]), re-identification rates can be high using methods originally developed for assessing re-identification risk for micro-aggregation (see Sections 18.4.6 and 18.4.7). While detailed discussion of the re-identification risk is beyond the scope of this text, we note that we can *manually* re-identify using our knowledge of how a rank-swapping file is produced.

18.4.4. Rounding

Rounding is a masking procedure that blurs the data by limiting the possible values a variable can take on. In particular, we define *the rounded value of x*

TABLE 18.4. Sample pairwise correlation coefficients based on original data

Variables	Variables			
	X_1	X_2	X_3	X_4
X_1	1.000	0.895	0.186	−0.102
X_2	0.895	1.000	0.218	−0.065
X_3	0.186	0.218	1.000	0.329
X_4	−0.102	−0.065	0.329	1.000

TABLE 18.5. Sample pairwise correlation coefficients based on a 5% rank swap

Variables	Variables			
	X_1	X_2	X_3	X_4
X_1	1.000	0.883	0.199	−0.063
X_2	0.883	1.000	0.206	−0.060
X_3	0.199	0.206	1.000	0.366
X_4	−0.063	−0.060	0.366	1.000

TABLE 18.6. Sample pairwise correlation coefficients based on a 10% rank swap

Variables	Variables			
	X_1	X_2	X_3	X_4
X_1	1.000	0.854	0.171	−0.077
X_2	0.854	1.000	0.169	−0.052
X_3	0.171	0.169	1.000	0.364
X_4	−0.077	−0.052	0.364	1.000

TABLE 18.7. Sample pairwise correlation coefficients based on a 20% rank swap

Variables	Variables			
	X_1	X_2	X_3	X_4
X_1	1.000	0.733	0.121	−0.163
X_2	0.733	1.000	0.152	−0.077
X_3	0.121	0.152	1.000	0.306
X_4	−0.163	−0.077	0.306	1.000

to the base b to be equal to $b \times \lfloor \frac{x}{b} + \frac{1}{2} \rfloor$ where $\lfloor y \rfloor$ is defined to be the largest integer $\leq y$. We illustrate this concept in Example 18.11.

Example 18.11

What is the rounded value of 172 to the base 50? The answer is

$$ b \times \left\lfloor \frac{x}{b} + \frac{1}{2} \right\rfloor = 50 \times \left\lfloor \frac{172}{50} + \frac{1}{2} \right\rfloor = 50 \times \left\lfloor \frac{197}{50} \right\rfloor = 50 \times 3 = 150. $$

Example 18.12

In this example, we again begin with the original dataset of Example 18.8. We then perform three types of rounding – to base 10, 50, and 100. The results are summarized in Tables 18.8–18.11.

With rounding to base 100, variables X_1 and X_3 can only take on two possible values; X_2 can only take on three possible values; and X_4 can only take on five different values. Although easy to implement, we see that rounding significantly affects analytic properties. Although covering identification risk is beyond the scope of this work, the re-identification risk may be high with the first three levels of rounding.

18.4.5. Global Recoding of Values in Certain Fields

This involves reducing the size of a partitioning scheme. For instance, instead of partitioning the United States into 50 States, we might instead partition it into four

TABLE 18.8. Sample pairwise correlation coefficients based on original data

Variables	Variables			
	X_1	X_2	X_3	X_4
X_1	1.000	0.895	0.186	−0.102
X_2	0.895	1.000	0.218	−0.065
X_3	0.186	0.218	1.000	0.329
X_4	−0.102	−0.065	0.329	1.000

TABLE 18.9. Sample pairwise correlation coefficients based on original data rounded to base 10

Variables	Variables			
	X_1	X_2	X_3	X_4
X_1	1.000	0.881	0.170	−0.115
X_2	0.881	1.000	0.205	−0.062
X_3	0.170	0.205	1.000	0.315
X_4	−0.115	−0.062	0.315	1.000

TABLE 18.10. Sample pairwise correlation coefficients based on original data rounded to base 50

Variables	Variables			
	X_1	X_2	X_3	X_4
X_1	1.000	0.739	0.096	−0.173
X_2	0.739	1.000	0.183	−0.107
X_3	0.096	0.183	1.000	0.289
X_4	−0.173	−0.107	0.289	1.000

TABLE 18.11. Sample pairwise correlation coefficients based on original data rounded to base 100

Variables	Variables			
	X_1	X_2	X_3	X_4
X_1	1.000	0.688	0.141	0.025
X_2	0.688	1.000	0.092	0.057
X_3	0.141	0.092	1.000	0.320
X_4	0.025	0.057	0.320	1.000

regions. Also, instead of partitioning the age field by individual ages, we could partition it into 5-year intervals. Efficient algorithms (Bayardo and Agrawal [2005], Lefevre et al. [2005]) for global recoding to achieve k-anonymity currently exist. A file is *k-anonymous* if every record $r \in \mathbf{Z}$ has *k-1* other records in $\mathbf{Z}$ that are identical to it. The analytic properties of global recoding (and k-anonymity) are currently unclear and are currently ongoing research problems.

18.4.6. Univariate k-Micro-aggregation

Some researchers have suggested sorting the values of individual fields and then partitioning the sorted vector into groups of k consecutive values each. The value of each element of the group is then replaced by an aggregate such as the average or median of the k values. Typically, a different k-partition is obtained for each variable. This is referred to as *univariate k-micro-aggregation*. Usually, k is taken to be 3 or 4 because simple analytic properties such as correlations degrade rapidly as k increases. Here k is a constant value chosen in advance by the data protector.

A more sophisticated approach to this problem is to use operations research techniques to obtain an "optimal" partition of the values of each individual field of the database where the size of each group is allowed to vary from k to $2k$-1. This is discussed in detail in Domingo-Ferrer and Mateo-Sanz [2002].

Some researchers have examined the effect of univariate micro-aggregation on simple regression statistics (Defays and Nanopolous [1992], Domingo-Ferrer and Mateo-Sanz [2002]). They found that if k is small (say $k = 3$ or $k = 4$), then the deterioration of regression coefficients would be minor. In the small k situation (and not-so-small k situation), Winkler (2002) demonstrated that new metrics could be quickly built into re-identification software that allows exceptionally high re-identification rates. At least one other researcher (Muralidhar [2003]) has verified Winkler's findings.

18.4.7. Multivariable k-Micro-aggregation

Multivariate k-micro-aggregation refers to techniques that allow simultaneous micro-aggregation of two or more variables so that the result is a single k-partition of the entire database. Domingo-Ferrer and Mateo-Sanz [2002] suggest using several (i.e., $n \geq 2$) fields at a time in a clustering procedure in which the database is partitioned into groups of size k to 2k-1 records. Here the values on the fields of the records within each cluster are replaced by an aggregate, such as those of the centroid of the cluster. Again operations research methods are employed to obtain an "optimal" partition. As they suggest and can be easily checked, such an n-field, k-aggregation can quickly degrade simple statistics such as correlation coefficients. We consider such an example next.

Example 18.13

The file, $\mathbf{X}$, consists of 100 records each of which has four variables (Tables 18.12 and 18.13). The file is somewhat similar to the artificial data used for the previous

TABLE 18.12. Sample pairwise correlation coefficients based on original data, **X**

Variables	Variables			
	X_1	X_2	X_3	X_4
X_1	1.000	−0.274	0.316	0.273
X_2	−0.274	1.000	0.044	0.039
X_3	0.316	0.044	1.000	0.935
X_4	0.273	0.039	0.935	1.000

TABLE 18.13. Sample pairwise correlation coefficients based on 10-micro-aggregated data, **Y**

Variables	Variables			
	X_1	X_2	X_3	X_4
X_1	1.000	−0.300	0.347	−0.361
X_2	−0.300	1.000	0.041	0.096
X_3	0.347	0.041	1.000	−0.305
X_4	−0.361	0.096	−0.305	1.000

empirical examples. To produce masked data **Y**, the data are partitioned into 10 clusters of 10 records each and each record in a cluster is replaced by its centroid (average of the 10 records).

It is clear that, at least in some situations, multivariable k-micro-aggregation yields severe deterioration in some statistics.

18.4.8. Further Remarks

Rank swapping with small p can approximately preserve some pairwise correlations. Swapping, on the other hand, almost instantly destroys all pairwise correlations.

If a statistical agency needs to produce a public-use file, then the agency should first consider the analytic properties that might be needed in the masked data **Z**. The optimal situation is when the users of the data all have identical and very narrowly focused analytic needs – for example, only one set of analysis.

18.5. Protecting Confidentiality of Medical Data

The medical and pharmaceutical industries are two of the most data intensive industries in the United States. Every day they create millions of records related to prescription drugs, lab tests, hospital treatment, and physician treatment. These records are valuable to pharmaceutical companies whose understanding of patients' and physicians' behavior over time is used in large part to formulate their marketing strategies. For example, by knowing (1) how long patients take a certain drug and in what quantities/dosage levels, (2) which competing drugs patients switch to and from, and (3) which doctors are most likely to prescribe a particular drug, pharmaceutical companies can develop targeted sales strategies to position themselves better in the marketplace.

In the United States, the Health Insurance Portability and Accountability Act of 1996 imposes strong restrictions on the disclosure of protected health information by healthcare providers and health insurers. Typically, entities covered by HIPAA must ensure that their data are de-identified before such data can be given to an outside entity, for example, an outside marketing firm. Therefore, as discussed earlier in this chapter, direct identifiers such as names, Social Security numbers, and home addresses as well as specific information on birth dates and ZIP codes must be removed. At the same time, market researchers still need to link records from different sources and over time, even if the records are de-identified. This creates a challenge because linkage is allowed across datasets – for example, those containing (1) lab test results and (2) prescription drug records – but linkage is not allowed between these de-identified datasets and an outside dataset that could be used for re-identification.

One way to address this issue is to create anonymous linking codes during the de-identification and production of the public-use file. Before being removed, the identifying information (name, address, birth date, gender, etc.) is used to create an essentially unique code for each individual. This code can be used internally to link records among internal datasets, but cannot be used to link the data to external, identified datasets.

In order to make this work, the linking codes must be designed so that an intruder cannot reasonably reverse-engineer them, that is recover the identifying information used to create the linking code. This can only be done using sophisticated cryptographic tools. The most common approach is to (1) concatenate all identifying fields into one string of data, (2) standardize the data (e.g., remove all special characters, make all letters upper case, and standardize dates), and (3) process this standardized string through a cryptographic "hash" function – a one-way function designed to make it hard for an intruder to take the output and back-step through the algorithm to recover the input.

Even though such hash algorithms are widely available and cryptographically sophisticated, there are significant challenges in the implementation of such algorithms. For example, if it is necessary to incorporate a secret key into the algorithm to prevent an intruder from recalculating the linking codes for known individuals, then the protection of the secret key is crucial.

HIPAA is an example how statisticians are increasingly brought in contact with other disciplines, such as cryptography, to meet the ever more sophisticated needs of de-identification. The cryptographic approach to de-identification and record linkage is an active research area with much work left to do. Baier [2007], Bournazian [2007] and Zarate [2007] recently presented work in this area.

18.6. More-Advanced Masking Methods – Synthetic Datasets

None of the models discussed earlier in this chapter involves concocting any data. In this section, we summarize[2] a number of procedures for constructing public-use datasets that entail the creation of synthetic data. In some instances only a few data values are synthetic. Under other schemes the entire dataset is synthetic.

We provide a number of references to the literature for those readers desiring more details on these methods. We begin with some of the procedures that are among the easiest to implement.

18.6.1. Kennickell [1999], Abowd and Woodcock [2002, 2004] and Dandekar et al. [2002]

Kennickell [1999] extended multiple imputation ideas to the generation of "synthetic" micro-data by successively blanking values in fields and filling in the data according to a (multiple-imputation) model. The ideas have also been effectively applied and extended by Abowd and Woodcock [2002, 2004] who have made substantial effort to demonstrate the analytic properties that are preserved and some of the limitations of the resulting file. The issue of whether the Kennickell type of methods allow re-identification is still open, although Dandekar et al. [2002] demonstrated that synthetic data generated via a "Latin Hypercube Sampling Technique" could preserve some analytic properties while providing disclosure protection.

18.6.2. Kim and Winkler [1995] – Masking Micro-data Files

Kim and Winkler [1995] proposed a procedure that they called *controlled distortion*. Under this procedure, Kim and Winkler [1995] proposed that the data provider (1) change easily identifiable records (e.g., outliers) in an arbitrary fashion and (2) make a series of offsetting changes on other records so that means and variances are preserved within subdomains. Kim and Winkler [1995] suggested moving easily identified outliers (specific values of certain fields on certain records) into the interiors of distributions while preserving the first two moments of the data. For every outlier whose values are changed, a (possibly large) number of values in fields of other records may need to be changed. An advantage of controlled distortion is that it preserves the confidentiality of the data. One disadvantage of controlled distortion is that means and correlations are

---

[2] The discussion in this section is tied closely to original sources. In some cases, we have quoted extensively from these sources. In others, we have closely paraphrased the discussion.

not usually preserved across subdomains. Another disadvantage is that if there are not sufficient records in a subdomain where moments need to be preserved, then the set of equations associated with the moments cannot be preserved.

18.6.3. Reiter – Satisfying Disclosure Restrictions with Synthetic Datasets

Rubin [1993] advocated creation of "multiple, synthetic datasets for public release so that (1) no unit in the released data has sensitive data from an actual unit in the population and (2) statistical procedures that are valid for the original data are valid for the released data." Reiter [2002] extends the work of Rubin [1993]. Specifically, Reiter [2002] "show[s] through simulation studies that valid inferences can be obtained from synthetic data in a variety of settings including simple random sampling, probability proportional to size sampling, two-stage cluster sampling, and stratified sampling. [Reiter] also provide[s] guidance on specifying the number and size of synthetic data sets and demonstrate[s] the benefit of including design variables in the released data sets."

18.6.4. Multiple Imputation for Statistical Disclosure Limitation

Raghunathan, Reiter, and Rubin [2003] evaluate "the use of the multiple imputation framework to protect the confidentiality of respondents' answers in sample surveys. The basic proposal is to simulate multiple copies of the population from which these respondents have been selected and release a random sample from each of these synthetic populations. Users can analyze the synthetic sample data sets with standard complete data software for simple random samples, then obtain valid inferences by combining the point and variance estimates using the methods in this article. Both parametric and non-parametric approaches for simulating these synthetic databases are discussed and evaluated. It is shown, using actual and simulated data sets in simple settings, that statistical inferences from these simulated research databases and the actual data sets are similar, at least for a class of analyses. Arguably, this class will be large enough for many users of public-use files. Users with more detailed demands may have to apply for special access to the confidential data."

The next procedure, while not designed to address the issue of statistical disclosure limitation, may nevertheless be of interest to those researching that issue.

18.6.5. DuMouchel et al. [1999] – "Squashing Flat Files Flatter"

The authors/researchers propose a scheme for summarizing a large dataset by a smaller synthetic dataset (frequently, several orders of magnitude smaller). The squashed dataset

* contains the same variables as the original dataset
* preserves low-order moments

- is constructed to emulate the multivariate distribution of the original dataset more accurately than would a dataset obtained by use of random (probability) sampling
- has the same structure as the original dataset with the addition of a *weight* for each pseudo-data point.

Any statistical modeling or machine learning method (software) that accepts weights can be used to analyze the pseudo-data. By construction, the resulting analyses will mimic the corresponding analyses on the original dataset.

DuMouchel et al. [1999] show how to create a set of aggregates consisting of a large number of the moments needed to approximate likelihoods used to create particular types of models for a set of continuous data. Moore and Lee [1998] show how to create a large number of approximations of aggregates needed for the loglinear modeling of discrete data. In each of these situations, the authors are interested in taking a file consisting of millions of records and providing a large set of records corresponding roughly to the record that could be obtained from the original data. A file of millions of records might be replaced by a file of thousands or tens of thousands of aggregated "records." The aggregated information or "records" would approximately reproduce key aggregates in the original file. In our context, we are only interested in how the analytic constraints affect the forms that the micro-data can assume in order that the masked micro-data be close to the original micro-data.

Because key insights provided by DuMouchel et al. [1999] are representative of similar insights provided by Moore and Lee [1998], we only describe the DuMouchel et al. work.

Although the procedure of DuMouchel et al. [1999] consists of three components, we are only concerned with their procedure for maintaining moments on subdomains. DuMouchel et al. [1999] are concerned with the log-likelihoods

$$\sum_{i=1}^{M} w_i \log \left[f \left(B_{i1}, \ldots, B_{iC}, Y_{i1}, \ldots, Y_{iQ}; \theta \right) \right] =$$

$$\sum_{i=1}^{N} \log \left[f \left(A_{i,1}, \ldots A_{iC}, X_{i1}, \ldots, X_{iQ}; \theta \right) \right]$$

(18.1)

Here f is the likelihood, the B's and A's are categorical variables, and the Y's and X's are quantitative variables. The B's and Y's are associated with artificial (synthetic) data that might preserve certain aggregates needed to approximate the log-likelihoods. In our situation, we have no categorical variables – that is we do not have any B's or A's. DuMouchel et al. assume that the log-likelihoods are sufficiently smooth functions that they can be approximated by moments of the quantitative variables – the Y's and X's. The original data might have $N = 1,000,000$ records that we wish to approximate with $M = 10,000$ or $M = 100,000$ records. It was their purpose to create data Y by solving a system of equations corresponding to (18.1) and satisfying a specified set of moment equations. The new data Y would be suitable for use in data mining or statistical software packages that can handle data Y with weights w_i where the original data X was too large for most software.

As DuMouchel et al. [1999] show, each record in $\mathbf{Y}$ may represent multiple records in $\mathbf{X}$. The degrees of freedom in solving the equations are represented by the number of records, M, in dataset $\mathbf{Y}$. As the number of moment constraints increases, the reduction factor, F, in going from N to M can decrease from a factor of $F >> 1$ to a factor of $F = 1$. In other words, as the number of moment constraints increases, the size, M, of dataset $\mathbf{Y}$ can increase until it reaches N – the number of records in the original dataset, $\mathbf{X}$.

Superficially, the procedure of DuMouchel et al. [1999] is similar to Kim and Winkler's [1995] controlled distortion procedure. The number of records and the number of variables approximately represent the number of degrees of freedom in the computation.

As an alternative to the DuMouchel et al. [1999] methods, Owen [2003] observed that, in some situations, it is more straightforward to draw a random sample and use a statistical procedure known as "empirical likelihood" to preserve some of the analytic constraints. Whereas the DuMouchel et al. methods may require special programming, there is high-quality generalized software for empirical likelihood.

18.6.6. Other Thoughts

Subdomains: Dandekar's work and earlier work on additive noise very strongly indicate that if analytic properties need to be preserved on subdomains, then the modeling needs to be done on the individual subdomains. There are two requirements on the subdomain modeling. First, the subdomains must partition the entire set of records. Second, each subdomain must be reasonably large. For very simple analytic properties, the subdomain will likely need a thousand or more records. For a moderate number of aggregates, the subdomain may need many tens of thousands of records (DuMouchel et al. [1999], Kim and Winkler [1995]). If any analyses need to be performed on the set of subdomains and on the entire file, then each record must satisfy two sets of analytic restraints. In the simple case of regression (which also preserves correlations), two sets of moment conditions need to be satisfied by all of the data records (with either additive-noise-masked data or synthetic data).

Reaction of Public Users: Will potential users of the "synthetic" dataset $\mathbf{Y}$ raise any objections to such artificial data?

Accuracy of Synthetic Data: How close to the original dataset $\mathbf{X}$ must the synthetic dataset $\mathbf{Y}$ be to allow reproduction of one or two analyses on the original dataset $\mathbf{X}$?

18.7. Where Are We Now?

In this chapter we have provided background and examples related to some of the common masking methods used to produce public-use micro-data. We demonstrated that elementary, easy-to-implement methods cause severe deterioration of analytic properties, usually without minimizing the danger of re-identification. More advanced masking methods based on density estimation (Raghunathan et al. [2003], Reiter [2003a, 2003b, 2005], systematic iterative blank-and-impute according to a model (Kennickel [1999], Abowd and Woodcock [2002, 2004]), or linearly transformed noise (Kim [1986, 1990]; Yancey et al. [2002]), are then introduced; these are known to preserve some analytic properties in masked micro-data. For each method, the paper we reference (1) describes how the masked data **Y** should be analyzed and (2) points out limitations on the use of the masked data **Y** for other analyses.

Appendix to Chapter 18

The benefits derived from record linkage projects can be substantial both in terms of dollars saved and the timeliness of the results. The process of linking records on individuals is intrinsically privacy intrusive, in the sense that information is brought together about a person without his or her knowledge or control. So, it should not be surprising that when data are used in record linkage studies for purposes beyond which they were specifically obtained, concerns may arise regarding the privacy of those data. Privacy concerns and other legal obligations need to be met in every record linkage study.

With these issues in mind, Statistics Canada (see Fellegi [1997]) has promulgated its privacy policy – which it hopes is both "cautious" and "balanced" – in regard to record linkage as follows.

Statistics Canada will undertake record linkage activities only if all of the following conditions are satisfied:

- The purpose of the record linkage activity is statistical/research and is consistent with the mandate of Statistics Canada as described in the *Statistics Act*; and
- The products of the record linkage activity will be released only in accordance with the confidentiality provisions of the *Statistics Act* and with any applicable provisions of the *Privacy Act*; and
- The record linkage activity has demonstrable cost or respondent burden savings over other alternatives, or is the only feasible option; and
- The record linkage activity will not be used for purposes that can be detrimental to the individuals involved and the benefits to be derived from such a linkage are clearly in the public interest; and
- The record linkage activity is judged not to jeopardize future conduct of Statistics Canada's programs; and
- The linkage satisfies a prescribed review and approval process.

19
Review of Record Linkage Software

From the 1950s through the early 1980s, researchers and practitioners undertaking a large record linkage project had no choice but to develop their own software. They often faced the choice of using less accurate methods or expending dozens of staff years to create proprietary systems. For example, in the late 1970s, the US National Agricultural Statistics Service spent what is conservatively estimated as 50 staff years to develop a state-of-the-art system. Happily, today's record linkage researchers and practitioners no longer need to do this any more. Powerful, flexible, relatively inexpensive software that implements all but the most sophisticated methods is available in the form of (1) generalized packages that can stand alone or (2) software components that can be integrated into a surrounding application.

19.1. Government

19.1.1. Statistics Canada – GRLS and Match360

Statistics Canada has developed the Generalized Record Linkage System (GRLS) and Match360. GRLS was developed to link records in the absence of unique identifiers. It is based on statistical decision theory. GRLS incorporates both the NYSIIS and the SOUNDEX phonetic coding systems. GRLS is designed to handle one-file (internal) and two-file record linkages. Possible applications include

- unduplicating mailing address lists (one-file)
- bringing hospital admission records together to construct "case histories"
- linking a file of workers exposed to potential health hazards to a mortality database (two-file)[1] and other epidemiology studies.

---

[1] The purpose here might be to detect the health risks associated with a particular occupation.

19.1.2.   US Census Bureau – GDRIVER

This software parses name and address strings into their individual components and presents them in a standard format. Several reference files are used to assist in this process. Some of these references contain lists of tokens that influence the way parsing is carried out. For example, if a conjunction token is found, the software must then consider producing more than one standardized record.

19.2.   Commercial

19.2.1.   AutoMatch and AutoStan – IBM

AutoMatch and AutoScan were both originally developed by Matthew Jaro and are now under the purview of IBM.

The AutoMatch Generalized Record Linkage System is a state-of-the-art software implementation of probabilistic record linkage methodology for matching records under conditions of uncertainty. AutoMatch simulates the thought process a human being might follow while examining and identifying data records representing a common entity or event. AutoMatch's comparative algorithms manage a comprehensive range of data anomalies and utilize frequency analysis methodology to precisely discriminate weight score values.

The AutoStan Generalized Standardization System is an intelligent pattern recognition parsing system that conditions records into normalized/standardized fix fielded format. AutoStan optimizes the performance of any linkage or matching system that utilizes consumer or business names and/or address data as identifiers during a match comparison.

19.2.2.   Trillium

The Trillium Software System is a set of tools for investigating, standardizing, enriching, linking and integrating data. This system can uncover previously untracked information in business records that is latent due to data entry errors, inconsistencies, unrecognized formats, and/or omissions. The Trillium System can be used to append geographical, census, corporate, and/or third-party infor-mation to enhance existing databases. Finally, it has record linkage software to perform probabilistic matching.

19.2.3.   APL as a Development Tool

With APL it is important to (1) avoid cramming, (2) use lots of comment statements, (3) clean-up code after it is up and running, and (4) realize that the person documenting the code may well be the next person using the code – be it in 1 month, 6 months, or 3 years.

19.3.   Checklist for Evaluating Record Linkage Software[2]

Day [1995] has created a checklist as an aid in evaluating commercial record linkage software. Day [1995] points out that while his list is long, there may still be important items specific to individual applications that he has omitted from his list. He adds that there is no substitute for a thorough analysis of one's individual needs.

19.3.1.   General

1. Is the software a generalized system or specific to a given application?
2. What is the form of the software?
   A complete system, ready to perform linkages right "out of the box"?
   A set of components, requiring that a system be built around them? If so, how complete are the components?
   A part of a larger system for performing data integration and cleaning functions?
3. Which types of linkage does the software support?
   Unduplication (one file linked to itself)?
   Linking two files?
   Simultaneously linking multiple files?
   Linking one or more files to a reference file (e.g., geographic coding)?
4. On which of the following types of computers will the software operate?
   Mainframe
   Workstation/Personal Computer
   Apple
   Parallel Cluster
5. On which of the following operating systems will the software run?
   Windows XP
   UNIX/Linux/Solaris
   VMS
   OSX
   Mainframe OS (e.g., IBM MVS)
6. For PC-based systems, what is the minimum size of (1) the CPU, (2) RAM, and (3) hard-drive?
7. Can the system perform linkages interactively (in real time)? Can it operate in batch mode?
8. How fast does the software run on the user's hardware given the expected size of the user's files? If the software is interactive, does it run at an acceptable speed?
9. If the software is to be used as part of a statistical analysis system, are the methods used in the software statistically defensible?

---

[2] This section, excluding the introductory paragraph, is written by Charles Day.

10. Is the software reliable? Can the vendor provide adequate technical support? Is the vendor expected to remain in business through the projected life of the software? If there is doubt about this, is a software escrow available? Is the user prepared to support the software himself/herself?
11. How well is the software documented?
12. What additional features does the vendor plan to make available in the near future (i.e., in the next version of the software)?
13. Is there a user group? Who else is using the software? What features would they like to see added? Have they developed any custom solutions (e.g., front ends, comparison functions) that they would be willing to share?
14. Is other software, such as a database package or editor needed to run the system?
15. Does the software contain security and data integrity protection features?
16. How many and what type of staff personnel will be required to develop a system from the software? To run the system? What type of training will they need? Will the vendor provide such training?

19.3.2. Record Linkage Methodology

1. Is the record linkage methodology based on (1) Fellegi–Sunter, (2) information-theoretic methods, and/or (3) Bayesian techniques?
2. How much control does the user have over the linkage process? Is the system a "black box" or can the user set parameters to control the record linkage process?
3. Does the software require any parameter files? If so, is there a utility provided for generating these files? How effectively does it automate the process? Can the utility be customized?
4. Does the user specify the linking variables and the types of comparisons?
5. What kinds of comparison functions are available for different types of variables? Do the methods give proportional weights (i.e., allow degrees of agreement)?
6. Can the user specify critical variables that must agree for a link to take place?
7. How does the system handle missing values for linkage variables? Does the system

   - compute a weight just as it does for any other value?
   - use a median between agreement and disagreement weights?
   - use a weight of zero?
   - allow the user to specify the desired approach?

8. Does the system allow array-valued variables (e.g., multiple values for phone number)? How do array-valued comparisons work?
9. What is the maximum number of linking variables?

10. Does the software support multiple linkage passes with different blocking and different linkage variables?
11. How does the software block variables? Do users set blocking variables? Can a pass be blocked on more than one variable?
12. Does the software contain or support routines for estimating linkage errors?

19.3.3.   Fellegi–Sunter Systems

1. How does the system determine the m- and u-probabilities? Can the user set m- and u-probabilities? Does the software contain utilities that set the m- and u-probabilities?
2. How does the system determine the weight cut-offs? Can the user set these? Does the software contain any utilities for determining the weight cut-offs?
3. Does the software permit the user to specify (1) the linkage weights and/or (2) the weights for missing values?

19.3.4.   Data Management

1. In what file formats can the software use data – (1) flat file, (2) SAS dataset, and/or (3) database? If the answer is "yes" to item (3), what types of database can be employed: Fox Pro, Informix, Sybase, Oracle, MySQL, DBII?
2. What is the maximum file size (number of records) that the software can handle?
3. How does the software manage records? Does it use temporary files or sorted files? Does it use pointers? Does it take advantage of database indexing?
4. Can the user specify the subsets of the data to be linked?
5. Does the software provide for "test matches" of a few hundred records to test the specifications?
6. Does the software contain a utility for viewing and manipulating data records?

19.3.5.   Post-linkage Functions

1. Does the software contain a utility for review of possible links? If so, what kind of functionality is provided for? What kind of interface does the utility use – character-based or GUI? Does the utility allow for review between passes, or only at the end of the process? Can two or more people work on the record review simultaneously? Can records be "put aside" for review later? Is there any provision for adding comments to the reviewed records pairs in the form of hypertext? Can pairs of groups of records be updated? Can the user "back up" or restore possible links before committing to decisions?

Can a "master" record be created that combines values from two or more records for different fields?

2. Does the software provide for results of earlier linkages (particularly reviews of possible links) to be applied to the current linkage process?
3. Does the software contain a utility for generating reports on the linked, unlinked, duplicate, and possible linked records? Can the format of the report be customized? Is the report viewed in character mode or is the report review done in a graphical environment? Can the report be printed? If so, what type of printer is required?
4. Does the software contain a utility that extracts files of linked and unlinked records? Can the user specify the format of such extracts?
5. Does the software generate statistics for evaluating the linkage process? Can the user customize the statistics generated by the system?

19.3.6.   Standardization

1. Does the software permit the user to partition variables in order to maximize the use of the information contained in these variables? For example, can a telephone number be partitioned into its (1) area code, (2) exchange, and (3) last four digits?
2. Can the name and address standardization/parsing components be customized? Can different processes be applied to different files?
3. Does the address standardization conform to US Postal Service standards?
4. Does the standardization modify the original data fields, or does it append standardized fields to the original data record?
5. How well do the standardization components work on the types of names the user wishes to link? For example, does the standardizer work well with Hispanic name? What about Asian names?
6. How well do the standardization components work on the types of addresses (e.g., rural or foreign) that the user expects to encounter?

19.3.7.   Costs

1. What are (1) the purchase price and (2) the annual maintenance cost of (i) the basic software system, (ii) additional features (e.g., database packages), and (iii) new or upgraded computer hardware?
2. What is the cost of training staff to use the system?
3. What are the estimated personnel and (in the case of mainframe systems) computer-time costs associated with running the system?
4. Is the cost of developing a new system for the intended purpose using the software within the available budget?
5. What is the upgrade path for the software? What will upgrades cost?
6. What kind of maintenance/support agreements is available? What do they cost?

19.3.8. Empirical Testing

1. What levels of false match and false non-match rates can be expected with the system? Are these levels acceptable?
2. How much manual intervention (e.g., possible match review) will the system require?
3. How rapidly will typical match projects be completed using the system?

20
Summary Chapter

In this text, we emphasize a number of themes.

First, in Chapter 2, we strongly advocate prevention over detection and repair. Specifically, we encourage the analyst to find innovative ways to prevent bad data from entering his/her databases. One way to do this is to construct a battery of front-end edits using the types of edits described in Chapter 5. Other ways include improving the questionnaire design or the interview process, although both of these are outside the scope of our work.

Second, we suggest that in many sample surveys, too high a portion of the available resources is devoted to editing the data. If less money were spent on editing, then (1) the data might not be over-edited to little or no avail and (2) the extra money could be used to improve other facets of the survey.

In Chapter 4, we describe (1) metrics for measuring data quality (or lack of it) and (2) metrics for assessing the quality of lists/databases produced by merging two or more lists/databases. Such metrics provide a quantitative objective measure of the quality of the data.

In Chapter 7, we describe the Fellegi–Holt automated editing model, hot-deck and model-based imputation models, as well as the advantages of multiple imputations. We also describe a unified edit/imputation system.

In Chapter 8, we describe the Fellegi–Sunter record linkage model while in Chapter 9 we show how to estimate the parameters of this model. There is an art to building both edit/imputation models and record linkage models that is distinct from the science and can only be obtained from close contact with data. So, we strongly encourage such contact.

In Chapter 10, we describe standardization and parsing techniques. These are used to facilitate comparisons between addresses and names.

In Chapter 11, we describe the Soundex System of Names and the NYSIIS Phonetic Decoder. Both of these are phonetic coding systems that can be used to block data in record linkage applications.

In Chapter 12, we show how blocking strategies can be used in record linkage applications to limit the number of pairs that need to be compared while at the same time minimizing the number of matches that are missed.

In Chapter 13, we describe string comparator metrics. These can be used to account for minor data entry errors and thereby facilitate record linkage applications.

In Chapter 14, we show how we used both data quality and record linkage techniques to improve a database of FHA-insured single-family mortgages.

In Chapters 15–17, we describe a number of applications of record linkage to a wide variety of subject areas. These areas include Health, Highway Safety, Agriculture, Censuses, and Federal Pensions.

In Chapter 18, we describe schemes that maximize public access to micro-data while protecting respondent privacy. This is still a fertile area for future research.

In Chapter 19, we review a variety of record-linkage software products and present a checklist for evaluating such software.

Bibliography

Abowd, J.M. and S.D. Woodcock, "Disclosure limitation in longitudinal linked data," in (P. Doyle et al, eds) *Confidentiality, Disclosure, and Data Access*, Amsterdam, The Netherlands: North Holland, 2002.

Abowd, J.M. and S.D. Woodcock, "Multiply-imputing confidential characteristics and file links in longitudinal linked data," in (J. Domingo-Ferrer and V. Torra, eds), *Privacy in Statistical Databases 2004*, New York: Springer, 2004.

Andrewatha, H.G., *Introduction to the Study of Animal Populations*, Chicago: University of Chicago Press, 1961.

Ash, R.B., *Basic Probability Theory*, New York: John Wiley & Sons 1970.

Baier, P., "Statistical deidentification and the HIPAA rule," to appear during 2007 in *Proceedings of the 2006 Joint Statistical Meetings*, American Statistical Association.

Ball, P., W. Betts, F.J. Scheuren, J. Dudukovich, and J. Asher, *Killings and Refuge Flow in Kosovo, March – June 1999*, American Academy for the Advancement of Science, January 3, 2002.

Bankier, M., A-M. Houle, M. Luc, and P. Newcombe, "1996 Canadian census demographic variables imputation," *Proceedings of the 1997 Section on Survey Research Methods, American Statistical Association*, pp. 389–394, 1997.

Barcaroli, G. and M. Venturi, "An integrated system for edit and imputation of data: an application to the Italian labour force survey," *Proceedings of the 49-th Session of the International Statistical Institute*, Florence, Italy, September 1993.

Barcaroli, G. and M. Venturi, "DAISY (design analysis and imputation system): Structure, methodology and first applications," in (J. Kovar and L. Granquist, eds) *Statistical Data Editing*, Volume II, UN Economic Commission for Europe, (also at http://www.unece.org/stats/documents/1997/10/data_editing/7.e.pdf), pp. 40–51, 1997.

Barnett, V. and T. Lewis, *Outliers in Statistical Data*, Third Edition, NewYork: John Wiley & Sons, 1994.

Baum, L.E., T. Petri, G. Soules, and N Weiss, "A maximization technique occurring in statistical analysis of probabilistic functions in Markov chains," *The Annals of Mathematical Statistics*, 41 (1), pp. 164–171, 1970.

Bayardo, R. J. and R. Agrawal, "Data privacy through optimal K-anonymization," *IEEE 2005 International Conference on Data Engineering*, 2005.

Belin, T.R. and D.B. Rubin, "A method for calibrating false matches in record linkage," *Journal of the American Statistical Association*, 90, pp. 694–707, 1995.

Berthelot, J.-M. and M. Latouche, "Improving the efficiency of data collection: a generic respondent follow-up strategy for economic surveys" *Journal of Business and Economic Statistics*, 11 (4), pp. 417–424, 1993.

Bienias, J., D. Lassman, S. Scheleur and H. Hogan, "Improving outlier detection in two establishment surveys," ECE Work Session on Statistical Data Editing, Working Paper No. 15, Athens, 6–9 November 1995.

Bishop, Y.M.M., S.E. Fienberg, and P.W. Holland, *Discrete Multivariate Analysis*, Cambridge, MA: MIT Press, 1975.

Bourdeau, R. and B. Desjardins, *Dealing with Problems in Data Quality for the Measurement of Mortality at Advanced Ages in Canada*, Department of Demography, University of Montreal, January, 2002.

Bournazian, J., "The Federal statisticians' response to the HIPAA privacy rule," to appear during 2007 in *Proceedings of the 2006 Joint Statistical Meetings*, American Statistical Association.

Brackett, M.H., *Data Resource Quality*. Boston, MA: Addison-Wesley, 2000.

Brackstone, G., "How important is accuracy?", *Proceedings of Statistics Canada Symposium 2001, Achieving Data Quality in a Statistical Agency: a Methodological Approach*, 2001.

Broadbent, K., *Record Linkage III: Experience Using AUTOMATCH in a State Office Setting*. STB Research Report Number STB-96-02. Washington, DC: National Agricultural Statistics Service, USDA, October, 1996.

Bruni, R. and A. Sassano, *Logic and optimization techniques for an error free data collecting*, Dipatimento di Informatica e Sistemistica, Universita di Roma "La Spienza", 2001.

Bruni, R., A. Reale, and R. Torelli, "Optimization techniques for edit validation and data imputation," *Statistics Canada Symposium 2001*, Ottawa, Ontario, Canada, October 2001.

Chapman, D.G., *Some Properties of the Hypergeometric Distribution with Applications to Zoological Censuses*, University of California, University of California Publication in Statistics, 1951.

Cleveland, W.S., *The Elements of Graphing Data*, Summit, NJ: Hobart Press, 1994.

Coale, A.J. and F.F. Stephan, "The case of the Indians and the teen-age widows," *Journal of the American Statistical Association*, 57, pp. 338–347, 1962.

Cohen, W.W., P. Ravikumar, and S.E. Fienberg, "A comparison of string distance metrics for name-matching tasks," *Proceedings of the ACM Workshop on Data Cleaning, Record Linkage, and Object Identification*, Washington, DC, August 2003a.

Cohen, W.W., P. Ravikumar, and S.E. Fienberg, "A comparison of string distance metrics for matching names and addresses," *Proceedings of the Workshop on Information Integration on the Web at the International Joint Conference on Artificial Intelligence*, Acapulco, Mexico, August 2003b.

Dalenius, T. and S.P. Reiss, (1982), "Data-swapping: A technique for disclosure control," *Journal of Statistical Planning and Inference*, 6, pp. 73–85, 1982.

Dandekar, R., M. Cohen, and N. Kirkendal, "Sensitive microdata protection using Latin hypercube sampling technique," in (J. Domingo-Ferrer, ed.) *Inference Control in Statistical Databases*, New York: Springer, pp. 117–125, 2002.

Dasu, T. and T. Johnson, *Exploratory Data Analysis and Data Mining*, New York: John Wiley & Sons, 2003.

Day, C., *Record Linkage I: Evaluation of Commercially Available Record Linkage Software for Use in NASS*. STB Research Report Number STB-95-02. Washington, DC: National Agricultural Statistics Service, USDA, October, 1995.

Day, C., *Record Linkage II: Experience Using AUTOMATCH for Record Linkage in NASS*. STB Research Report Number STB-96-01. Washington, DC: National Agricultural Statistics Service, USDA, March, 1996.

Defays, D. and P. Nanopolis, "Panels of enterprises and confidentiality: the small aggregates method," in *Proceedings of the 1992 Symposium on Design and Analysis of Longitudinal Surveys*, pp. 195–204, 1993.

DeGuire, Y., "Postal address analysis," *Survey Methodology*, 14, pp. 317–325, 1988.

Deming, W.E. *Quality Productivity and Competitive Position*, Cambridge, MA: MIT Press, 1982.

Deming, W.E., *Out of the Crisis*. Cambridge: MIT Center for Advanced Engineering Study, 1986.

Deming, W.E., "On errors in surveys," excerpt *The American Statistician*, 60 (1), pp. 34–38, February, 2006.

Deming, W.E. and G.J. Gleser, "On the problem of matching lists by samples," *Journal of the American Statistical Association*, 54, pp. 403–415, 1959.

Dempster, A., D.B. Rubin, and N. Laird, "Maximum likelihood from incomplete data via the EM algorithm" (with discussion), *Journal of the Royal Statistical Society*, Series B, 39, pp. 1–39, 1977.

De Waal, T., "New developments in automatic editing and imputation at statistics Netherlands," UN Economic Commission for Europe Work Session on Statistical Data Editing, Cardiff, UK, October 2000 (also available at http://www.unece.org/ stats/documents/2000.10.sde.htm).

De Waal, T., "Solving the error localization problem by means of vertex generation," *Survey Methodology*, 29 (1), 71–79, 2003a.

De Waal, T., "A fast and simple algorithm for automatic editing of mixed data," *Journal of Official Statistics*, 19 (4), 383–402, 2003b.

De Waal, T., "Computational results with various error localization algorithms," UNECE Statistical Data Editing Worksession, Madrid, Spain (also at http://www.unece.org/stats/documents/2003/10/sde/wp.22.e.pdf), 2003c.

De Waal, T., *Processing of Erroneous and Unsafe Data*, ERIM Research in Management: Rotterdam, 2003d.

Dippo, S.C. and B. Sundgren, *The Role of Metadata in Statistics*, paper presented at International Conference on Establishment Surveys II, Buffalo, New York, 2000. Paper available at http://www.bls.gov/ore/pdf/st000040.pdf.

Domingo-Ferrer, J. and J.M. Mateo-Sanz, "Practical data-oriented microaggregation for statistical disclosure control," *IEEE Transactions on Knowledge and Data Engineering*, 14 (1), pp. 189–201, 2002.

Draper, L. and W.E. Winkler, "Balancing and ratio editing with the new SPEER system," *American Statistical Association Proceedings of the 1997 Section on Survey Research Methods*, pp. 570–575 (also at http://www.census.gov/srd/papers/pdf/rr97-5.pdf), 1997.

Drew, J.D., D.A. Royce, and A. van Baaren, "Address register research at statistics Canada," *Proceedings of the Section on Survey Research Methods*, American Statistical Association, pp. 252–259 (also at http://www.amstat.org/sections/srms/proceedings/papers/1989_042.pdf), 1989.

DuMouchel, W., C. Volinsky, T. Johnson, C. Cortes, and D. Pregibon, "Squashing flat files flatter," *Proceedings of the ACM Knowledge Discovery and Data Mining Conference*, (also at http://www.research.att.com/~volinsky/kddsquash.ps), pp. 6–15, 1999.

English, L.P., *Improving Data Warehouse and Business Information Quality: Methods for Reducing Costs and Increasing Profits*, New York: John Wiley & Sons, 1999.

Federal Committee on Statistical Methodology, Report on Exact and Statistical Matching Techniques, http://www.fcsm.gov/working-papers/wp5.html, 1980.

Fellegi, I.P., "Record linkage and public policy – a dynamic evolution," *Record Linkage Techniques*, Washington, DC, Federal Committee on Statistical Methodology, Office of Management and Budget, (also at http://www.fcsm.gov/working-papers/fellegi.pdf), 1997.

Fellegi, I.P. and D. Holt, "A systematic approach to automatic data editing," *Journal of the American Statistical Association*, 71, pp. 17–35, 1976.

Fellegi, I.P. and A.B. Sunter, "A theory for record linkage," *Journal of the American Statistical Association*, 64, pp. 1183–1210, 1969.

Ford, B.N., "An overview of hot deck procedures," *Incomplete Data in Sample Surveys, Volume 2: Theory and Annotated Bibliography* in (W.G Madow, I. Olkin, and D.B. Rubin, eds), New York: Academic Press, 1983.

Gallop, A.P., *Mortality at Advanced Ages in the United Kingdom*. United Kingdom Government Actuary's Department, London, 2002.

Garfinkel, R.S., A.S. Kunnathur, and G.E. Liepens, "Optimal imputation of erroneous data: categorical data, general edits," *Operations Research* 34, pp. 744–751, 1986.

Garcia, M. and K.J. Thompson, "Applying the generalized edit/imputation system AGGIES to the annual capital expenditures survey," *Proceedings of the International Conference on Establishment, Surveys, II*, pp. 777–789, 2000.

Granquist, L. and J.G. Kovar, "Editing of survey data: How much is enough?", in *Survey Measurement and Process Control* (L. Lyberg, P. Biemer, M. Collins, E. de Leeuw, C. Dippo, N. Schwarz, and D. Trewin, eds), New York: John Wiley & Sons, pp.415–435, 1997.

Greenberg, B.G. and R. Surdi, "A flexible and interactive edit and imputation system for ratio edits," SRD report RR-84/18, US Bureau of the Census, Washington, DC (also at http://www.census.gov/srd/papers/pdf/rr84-18.pdf), 1984.

Haberman, S.J., "Iterative scaling for log-linear model for frequency tables derived by indirect observation", *Proceedings of the Section on Statistical Computing*, American Statistical Association, pp. 45–50, 1975.

Haberman, S.J., *Analysis of Qualitative Data*, New York: Academic Press, 1979.

Haines, D.E., K.H. Pollock, and S.G. Pantula, "Population size and total estimation when sampling from incomplete list frames with heterogeneous probabilities," *Survey Methodology*, 26(2), pp. 121–129, 2000.

Haworth, M.F. and J. Martin, *Delivering and Measuring Data Quality in UK National Statistics*, Office for National Statistics, UK (also at www.fcsm.gov/01papers/Haworth.pdf), 2001.

Herzog, T.N. and G. Lord, *Applications of Monte Carlo Methods to Finance and Insurance*, Winsted, CT: ACTEX Publications, 2002.

Herzog, T.N. and D.B. Rubin, "Using multiple imputation to handle nonresponse in sample surveys," in *Incomplete Data in Sample Surveys, Volume 2: Theory and Bibliography* (W.G Madow, I. Olkin, and D.B. Rubin, eds), New York: Academic Press, 1983.

Hidiroglou, M.A. and J.M. Berthelot, "Statistical editing and imputation for periodic business surveys," *Survey Methodology*, 12 (1), pp. 73–83, 1986.

Hinkins, S.M., "Matrix sampling and the related imputation of corporate income tax returns", *Proceedings of the Section on Survey Research Methods, American Statistical Association*, pp. 427–433, 1983.

Hinkins, S.M., "Matrix sampling and the effects of using hot deck imputation," *Proceedings of the Section on Survey Research Methods, American Statistical Association*, pp. 415–420, 1984.

Hogg, R.V., J.W. McKean, and A.T. Craig, *Introduction to Mathematical Statistics*, Sixth Edition, Upper Saddle River, NJ: Prentice Hall PTR, 2005.

Huang, K.T., R.Y. Wang and Y.W. Lee, *Quality Information and Knowledge*, Upper Saddle River, NJ: Prentice Hall PTR, 1999.

Imai, Kaizen: *The Key to Japan's Competitive Success*. New York: McGraw-Hill, 1986.

Jaro, M.A., "UNIMATCH –a computer system for generalized record linkage under conditions of uncertainty," *Spring Joint Computer Conference, 1972, AFIPSL-Conference Proceedings*, 40, pp. 523–530, 1972.

Jaro, M.A., "Advances in record-linkage methodology as applied to matching the 1985 census of Tampa, Florida," *Journal of the American Statistical Association*, 84 (406), pp. 414–420, 1989.

Jaro, M.A., "Probabilistic linkage of large public health datafiles," *Statistics in Medicine*, 14, pp. 491–498, 1995.

Johnson, S., "Technical issues related to probabilistic linkage of population-based crash and injury data," *Record Linkage Techniques*, Washington, DC, Federal Committee on Statistical Methodology, Office of Management and Budget, (also at http://www.fcsm.gov/working-papers/sandrajohnson.pdf), 1997.

Juran, J.M. and A.M. Godfrey (eds), *Juran's Quality Handbook*, Fifth Edition, McGraw-Hill, 1999.

Kalton, G., "How important is accuracy?", *Proceedings of Statistics Canada Symposium 2001, Achieving Data Quality in a Statistical Agency: a Methodological Approach*, 2001.

Kennickell, A.B., "Multiple imputation and disclosure control: The case of the 1995 survey of consumer finances," in *Record Linkage Techniques 1997*, Washington, DC: National Academy Press, pp. 248–267 (available at http://www.fcsm.gov under Methodology reports), 1999.

Kestenbaum, B., "Probability linkage using social security administration files", *Proceedings of 1996 Annual Research Conference*, US Census Bureau, Suitland, MD, 1996.

Kestenbaum, B., *Discussion of Session 4 (Sparse Data), 2002 Symposium on Living to 100 and Beyond: Survival at Advanced* Ages, available at http://www.soa.org/ccm/content/research-publications/research-projects/living-to-100-and-beyond-survival-at-advanced-ages, 2003.

Kestenbaum, B. and R. Ferguson, *Mortality of Extreme Aged in the United States in the 1990s, Based on Improved Medicare Data*. Social Security Administration, Baltimore, MD, 2002.

Kilss, B. and F.J. Scheuren, "The 1973 CPS-IRS-SSA exact match study", *Social Security Bulletin*, 41 (10). Baltimore, MD: pp. 14–22, 1978.

Kim, J.J., "A method for limiting disclosure in microdata based on random noise and transformation," *Proceedings of the Section on Survey Research Methods*, American Statistical Association, pp. 370–374 (http://www.amstat.org/sections/SRMS/Proceedings/papers/1986_069.pdf), 1986.

Kim, J.J., "Subdomain estimation for the masked data," *Proceedings of the Section on Survey Research Methods*, American Statistical Association, pp. 456–461, 1990.

Kim, J. J. and W.E. Winkler, "Masking microdata files," *Proceedings of the Section on Survey Research Methods*, American Statistical Association, pp. 114–119, 1995.

Kingkade, W., *Discussion of Session 7, Part I, (Mortality at the Oldest Ages), 2002 Symposium on Living to 100 and Beyond: Survival at Advanced* Ages, available at http://www.soa.org/ccm/content/research-publications/research-projects/living-to-100-and-beyond-survival-at-advanced-ages, 2003.

Knuth, D.E., *The Art of Computer Programming*, Vol. 3, Second Edition, Reading, PA: Addison-Wesley, 1998.

Kovar, J.G. and W.E. Winkler, "Editing economic data," *Proceedings of the Section on Survey Research Methods*, American Statistical Association, pp. 81–87, (also at http://www.census.gov/srd/papers/pdf/rr2000-04.pdf), 1996.

Kovar, J.G. and W.E. Winkler, *Comparison of GEIS and SPEER for Editing Economic Data*, Statistical Research Report Series, No. RR2000/04, US Bureau of the Census, Washington, DC, October 3, 2000.

Kovar, J.G., J.H. Macmillan, and P. Whitridge, "Overview and strategy for generalized edit and imputation system," Statistics Canada, Methodology Branch Working Paper BSMD 88-007E (updated during 1991), 1991.

Latouche, M. and J.-M. Berthelot, "Use of a score function to prioritize and limit recontacts in editing business surveys," *Journal of Official Statistics*, 8 (3), pp. 389–400, 1992.

Lawrence, D. and R McKenzie, "The general application of significance editing," *Journal of Official Statistics*, 16(3), pp. 243–253, 2000.

LeFevre, K., D. DeWitt, and R. Ramakrishnan, "Incognito: Efficient full-domain K-anonymity," *ACM SIGMOD Conference*, pp. 49–60, 2005.

Lincoln, F.C., "Calculating waterfowl abundance on the basis of banding returns," *Cir. U.S. Department of Agriculture*, 118, pp. 1–4, 1930.

Little, R.J.A and D.B. Rubin, *Statistical Analysis with Missing Data*. New York: John Wiley & Sons, 1987.

Little, R.J.A and D.B. Rubin, *Statistical Analysis with Missing Data*. Second Edition, New York: John Wiley & Sons, 2002.

Loshin, D., *Enterprise Knowledge Management: The Quality Approach*, San Francisco: Morgan Kaufman, 2001.

Lu, R.-P., P. Weir and R. Emery, "Implementation of the graphical editing analysis query system," US Energy Information Administration, Science Applications International Corporation, paper presented at 1999 research conference of the Federal Committee on Statistical Methodology. This paper is available at http://www.fcsm.gov/99papers/fcsmalldgr.html.

Lynch, B.T. and W.L. Arends, *Selection of a Surname Coding Procedure for the SRS Record Linkage System*, Sample Survey Branch, Research Division, Statistical Reporting Service, US Department of Agriculture, Washington, DC, February, 1977.

Marks, E.S., W. Seltzer, and K.R. Krotki, *Population Growth Estimation*, The Population Council, New York, 1974.

Meng, X.L. and D.B. Rubin, "Using EM to obtain asymptotic variance-covariance matrices: The SEM algorithm," *Journal of the American Statistical Association*, 86, pp. 899–909, 1991.

Meng, X.L. and D.B. Rubin, "Maximum likelihood via the ECM algorithm: A general framework," *Biomertika*, 80, pp. 267–278, 1993.

Moore, R., "Controlled data swapping techniques for masking public use data sets," US Bureau of the Census, Statistical Research Division Report rr96/04 (available at http://www.census.gov/srd/www/byyear.html), 1995.

Moore, A.W. and M.S. Lee, "Cached sufficient statistics for efficient machine learning with large datasets," *Journal of Artificial Intelligence Research*, 8, pp. 67–91, 1998.

Mosteller, F. and J.W. Tukey, *Data Analysis and Regression*, Reading, MA: Addison-Wesley Publishing Company, 1977.

Muralidhar, K., Private communication, 2003.

Mustaq, A. and F.J. Scheuren, *State Sales Tax Compliance Study, Volume I, Study Survey Summary*, NORC, University of Chicago, 2005.

Naus, J.I., *Data Quality Control and Editing*. New York: Marcel Dekker, 1975.

Neutel. C.I., "Privacy issues in research using record linkages," *Pharmcoepidemiology and Drug Safety*, 6, pp. 367–369, 1997.

Newcombe, H.B., *Handbook of Record Linkage: Methods for Health and Statistical Studies, Administration, and Business*. Oxford: Oxford University Press, 1988.

Newcombe, H.B. and J.M. Kennedy, "Record linkage: Making maximum use of the discriminating power of identifying information," *Communications of the Association for Computing Machinery*, 5, pp. 563–567, 1962.

Newcombe, H.B., J.M. Kennedy, S.J. Axford, and A.P. James, "Automatic linkage of vital records," *Science*, 130, pp. 954–959, 1959.

NIST, *Engineering Statistics Handbook*, Chapter 1, Exploratory Data Analysis (http://www.itl.nist.gov/div898/handbook/eda/eda.htm), 2005.

Nitz, L.H. and K.E. Kim, "Investigating auto injury treatment in a no-fault state: An analysis of linked crash and auto insurer data," *Record Linkage Techniques*, Washington, DC, Federal Committee on Statistical Methodology, Office of Management and Budget (also at http://www.fcsm.gov/working-papers/nitz-kim.pdf), 1997.

Office of Federal Statistical Policy and Standards, "Report on exact and statistical matching techniques," Statistical Policy Working Paper 5, US Department of Commerce (also at http://www.fcsm.gov/working-papers/wp5.html), 1980.

Oh, H.L. and F.J. Scheuren, "Estimating the variance impact of missing CPS income data", *Proceedings of the Section on Survey Research Methods, American Statistical Association* (also at http://www.fcsm.gov/working-papers/wp5.html), pp. 408–415, 1980.

Oh, H.L. and F.J. Scheuren, "Weighting adjustment for unit nonresponse," in *Incomplete Data in Sample Surveys, Volume 2: Theory and Bibliography* (W.G Madow, I. Olkin, and D.B. Rubin, eds), New York: Academic Press, 1983.

Oh, H.L., F.J. Scheuren, and H. Nisselson, "Differential bias impacts of alternative census bureau hot deck procedures for imputing missing CPS income data," *Proceedings of the Section on Survey Research Methods, American Statistical Association*, (also at http://www.amstat.org/Sections/Srms/Proceedings/papers/1980_086.pdf), pp. 416–420, 1980.

Owen, A., "Data squashing by empirical likelihood," *Data Mining and Knowledge Discovery*, 7 (1), pp. 101–113, 2003.

Petersen, C.G.J., "The yearly immigration of young plaice into the Limfiord from the German Sea," *Rep. Dan. Biol. Stn.*, 6, pp. 5–84, 1896.

Pipino, L.L., Y.W. Lee, and R.Y. Wang, "Data quality assessment," *Communications of the ACM*, 45 (4), pp. 211–218, 2002.

Pollock, J. and A. Zamora, "Automatic spelling correction in scientific and scholarly text," *Communications of the ACM*, 27, pp. 358–368, 1984.

Raghunathan, T.E., J.P. Reiter, and D.B. Rubin, "Multiple imputation for statistical disclosure limitation," *Journal of Official Statistics*, 19, pp. 1–16, 2003.

Redman, T.C., *Data Quality for the Information Age*. Boston: Artech House, 1996.

Reiter, J.P., "Inference for partially synthetic, public use data sets," *Survey Methodology*, pp. 181–189, 2003a.

Reiter, J.P., "Estimating probabilities of identification for microdata," Panel on Confidential Data Access for Research Purposes, Committee On National Statistics, October 2003, 2003b.

Reiter, J.P., "Releasing multiply imputed, synthetic public use microdata: An illustration and empirical study, *Journal of the Royal Statistical Society, A*, 169(2), pp. 185–205, 2005.

Rousseeuw, P. and A.M. Leroy, *Robust Regression and Outlier Detection*, New York: John Wiley & Sons, 2003.

Rubin, D.B., *Multiple Imputation for Nonresponse in Surveys*. New York: John Wiley & Sons, 1987.

Salazar-Gonzalez, J.-J. and J. Riera-Ledsema, "New algorithms for the editing-and-imputation problem," *UNECE Statistical Data Editing Worksession*, Madrid, Spain, 2003, http://www.unce.org/stats/documents/2003/10/sde/wp.5.e.pdf.

Sande, I.G., "Imputation in surveys: Coping with reality," *The American Statistician*, 36 (3), Part 1, 1982.

Scheuren, F.J., "How important is accuracy?", *Proceedings of Statistics Canada Symposium 2001, Achieving Data Quality in a Statistical Agency: a Methodological Approach*, 2001.

Scheuren, F.J., "The history corner," *The American Statistician*, 58 (1), 2004.

Scheuren, F.J., "Seven rules of thumb in interpreting nonresponse rates", *Presentation at NISS Nonsampling Error Conference*, March 17, 2005a. Available at www.niss.org/affiliates/totalsurveyerrorworkshop200503/presentations/Scheuren.pdf

Scheuren, F.J., "Multiple imputation: How it began and continues," *The American Statistician*, 59(4), 2005b.

Scheuren, F.J. and W.E. Winkler, "Regression analysis of data files that are computer matched," *Survey Methodology*, 19, pp. 39–58, 1993.

Scheuren, F.J. and W.E. Winkler, "Regression analysis of data files that are computer matched, II" *Survey Methodology*, 23, pp.157–165, 1997.

Sekar, C.C. and W.E. Deming, "On a method of estimating birth and death rates and the extent of registration," *Journal of the American Statistical Association*, 44, pp. 101–115, 1949.

Sekar, C.C. and W.E. Deming, "On a method of estimating birth and death rates and the extent of registration," (excerpt) *The American Statistician*, 58(1), pp. 13–15, February, 2004.

Smith, M.E. and H.B. Newcombe, "Methods for computer linkage of hospital admission-separation records into cumulative health histories," Meth Inform Med, 14, pp. 118–125, 1975.

Swain, L., Drew, J.D., Lafrance, B., and Lance, K., "The creation of a residential address register for use in the 1991 Census Survey," *Survey Methodology*, 18, pp. 127–141, 1992.

Taft, R.L., *Name Search Techniques*, New York State Identification and Intelligence System, Special Report No. 1, Albany, New York, 1970.

Thibaudeau, Y., "Fitting log-linear models when some dichotomous variables are unobservable," *Proceedings of the Section on Statistical Computing*, American Statistical Association, pp. 283–288, 1989.

Tukey, J.W., *Exploratoy Data Analysis*, Reading, MA: Addison-Wesley Publishing Company, 1977.

US General Accounting Office, *Record Linkage and Privacy: Issues in Creating New Federal Research and Statistical Information*, GAO-01-126SP, Washington, DC, April, 2001.

Utter, D., "Use of probabilistic linkage for an Analysis of the Effectiveness of Safety Belts and Helmets, in (W. Alvey and B. Kilss, eds) *Record Linkage Techniques 1997* (available at http://www.fcsm.gov/working-papers/utter.pdf), 1997.

Velleman, P.F. and D.C. Hoaglin, *Applications, Basics, and Computing of Exploratory Data Analysis*. Duxbury Press, Boston, Mass., 1981.

Wang, R.Y. "A product perspective on total data quality management," *Communications of the ACM*, 41 (2), pp 58–65, 1998.

Wang, R.Y., M. Ziad, and Y.W. Lee, *Data Quality*. 2001: Kluwer Academic Publishers, 2001.

Winkler, W.E., *Exact Matching Using Elementary Techniques*, technical report, Washington, DC: US Energy Information Administration, 1984.

Winkler, W.E., "Preprocessing of lists and string comparison," in (W. Alvey and B. Kilss, eds), *Record Linkage Techniques—1985*, US Internal Revenue Service. Publication 1299 (2-86), pp. 181–187 (also at http://www.fcsm.gov/working-papers/1367_3.pdf), 1985a.

Winkler, W.E., "Exact matching lists of businesses: Blocking, subfield identification, and information theory", in (W. Alvey and B. Kilss, eds), *Record Linkage Techniques—1985*, US Internal Revenue Service. Publication 1299 (2-86), pp. 227–241 (also at http://www.fcsm.gov/working-papers/1367_3.pdf), 1985b.

Winkler, W.F., "Using the EM algorithm for weight computation in the Fellegi–Sunter model of record linkage," *Proceedings of the Section on Survey Research Methods*, American Statistical Association, pp. 667–671 (also at http://www.census.gov/srd/papers/pdf/rr2000-05.pdf), 1988.

Winkler, W.E., "Methods for adjusting for lack of independence in an application of the Fellegi–Sunter model of record linkage," *Survey Methodology*, 15, pp. 101–117, 1989a.

Winkler, W.E., "Frequency-based matching in the Fellegi–Sunter model of record linkage," *Proceedings of the Section on Survey Research Methods, American Statistical Association*, pp. 778–783 (also at http://www.census.gov/srd/papers/pdf/rr2000-06.pdf), 1989b.

Winkler, W.E., "String comparator metrics and enhanced decision rules in the Fellegi–Sunter model of record," *Proceedings of the Section on Survey Research Methods*, American Statistical Association, pp. 354–359 (also at http://www.amstat.org/sections/srms/proceedings/papers/1990_056.pdf), 1990.

Winkler, W.E., "Error models for analysis of computer linked files," *Proceedings of the Section on Survey Research Methods*, American Statistical Association, pp. 472–477 (also at http://www.amstat.org/sections/srms/Proceedings/papers/1991_079.pdf), 1991.

Winkler, W.E., "Comparative analysis of record linkage decision rules," *Proceedings of the Section on Survey Research Methods*, American Statistical Association, pp. 829–834 (also at http://www.amstat.org/sections/srms/Proceedings/papers/1992_140.pdf), 1992.

Winkler, W.E., "Improved decision rules in the Fellegi–Sunter model of record linkage," *Proceedings of the Section on Survey Research Methods*, American Statistical Association, pp. 274–279 (also at http://www.census.gov/srd/papers/pdf/rr93-12.pdf), 1993.

Winkler, W.E., "Advanced methods for record linkage," *Proceedings of the Section on Survey Research Methods*, American Statistical Association, pp. 274–279 (also at http://www.census.gov/srd/papers/pdf/rr94-5.pdf), 1994.

Winkler, W.E., "Matching and record linkage," in (B.G. Cox, D.A. Binder, B.N. Chinnappa, M.J. Christianson, M.J. Colledge, and P.S. Kott, Eds), *Business Survey Methods*, New York, NY: John Wiley & Sons (also at http://www.fcsm.gov/working-papers/wwinkler.pdf), 1995.

Winkler, W.E., "The state of statistical editing," in *Statistical Data Editing*, Rome: ISTAT, pp. 169–187 (also at http://www.census.gov/srd/papers/pdf/rr99-01.pdf), 1999.

Winkler, W.E., "Single ranking micro-aggregation and re-identification," Statistical Research Division report RR 2002/08 (also at http://www.census.gov/srd/papers/pdf/rrs2002-08.pdf), 2002.

Winkler, W.F., "A contingency table model for imputing data satisfying analytic constraints," *Proceedings of the Section on Survey Research Methods, American Statistical Association* (also at http://www.census.gov/srd/papers/pdf/rrs2003-07.pdf), 2003.

Winkler, W.E., "Methods for evaluating and creating data quality," *Information Systems*, 29 (7), 531–550, 2004.

Winkler, W.E., "Approximate string comparator search strategies for very large administrative lists," *Proceedings of the Section on Survey Research Methods, American Statistical Association* (also at http://www.census.gov/srd/papers/pdf/rrs2005-02.pdf), CD-ROM, 2004.

Winkler, W.E. and T. Petkunas, "The DISCRETE edit system," in (J. Kovar and L. Granquist, eds) *Statistical Data Editing*, Vol. II, UN Economic Commission for Europe, pp. 56–62, 1997.

Winkler, W.E. and Y. Thibaudeau, *An Application of the Fellegi-Sunter Model of Record Linkage to the 1990 U.S. Census*, US Bureau of the Census, Statistical Research Division Technical Report (also at http://www.census.gov/srd/papers/pdf/rr91-9.pdf), 1991.

Winkler, W. E., "Record linkage course notes", Joint Program on Statistical Methodology, University of Maryland and University of Michigan, 1998.

Witte, G., "Your statements went where?", *The Washington Post*, February 6, 2005, p. F01.

Yancy, W.E. and W.E. Winkler, *Record Linkage Software: User Documentation*, Statistical Research Division, US Bureau of the Census, Suitland, MD. October 10, 2002.

Yancey, W.E., W.E. Winkler, and R.H. Creecy, "Disclosure risk assessment in perturbative microdata protection," in (J. Domingo-Ferrer, ed.) *Inference Control in Statistical Databases*, New York: Springer, pp. 135–151 (also http://www.census.gov/srd/papers/pdf/rrs2002-01.pdf), 2002.

Zarate. A.O., "The statistician's role in developing the HIPAA de-identification standard," to appear during 2007 in *Proceedings of the 2006 Joint Statistical Meetings*, American Statistical Association.

Zaslavsky, A.M. and G.S. Wolfgang, "Triple system modeling of census, post-enumeration survey, and administrative list data," *Journal of Business Economic Statistics*, 11, pp. 279–288, 1993.

Index

Printed in the United States of America